Modelling, Analysis and Optimization of Biosystems

Werner Krabs · Stefan Wolfgang Pickl

Modelling, Analysis and Optimization of Biosystems

Springer

Professor Dr. Werner Krabs
TU Darmstadt
Fachbereich Mathematik
Schloßgartenstraße 7
64289 Darmstadt
krabs@mathematik.tu-darmstadt.de

Professor Dr. Stefan Wolfgang Pickl
Universität der Bundeswehr München
Fakultät für Informatik
Werner-Heisenberg-Weg 39
85577 Neubiberg
stefan.pickl@unibw.de
http://www.unibw.de/stefan.pickl

Library of Congress Control Number: 2007931448

ISBN 978-3-540-71452-1 Springer Berlin Heidelberg New York

Springer is a part of Springer Science+Business Media

springer.com

Production: LE-TEX Jelonek, Schmidt & Vöckler GbR, Leipzig
Cover-design: WMX Design GmbH, Heidelberg

SPIN 12036585 42/3180YL - 5 4 3 2 1 0 Printed on acid-free paper

Preface

"Die rationale Entscheidungstheorie hat ihre Berechtigung. Sie ist ein wunderschönes mathematisches Modell einer vollkommenen Welt. Und sie erklärt auch einen Teil der Realität, zum Beispiel in der Biologie. Die Evolution erzeugt bei ganz und gar unvernünftigen Tieren und Pflanzen in gewisser Weise optimale Verhaltensprogramme, sie erzwingt bei einfachen Organismen gnadenlos Rationalität. Höhere Wesen wie wir, die viel stärker von der Kultur geprägt sind als von der Evolution, haben viel größeren Spielraum. Menschen verhalten sich nun mal nicht streng rational. Die Spieltheorie ist bestenfalls eine erste Annäherung an dieses Problem, eine Art idealer Maßstab. Das war mir schon Ende der fünfziger Jahre klar, als ich als Experimentalökonom anfing, theoretische Annahmen im Labor zu überprüfen."

Reinhard Selten

Mathematical models in biology and medicine cannot be based on natural laws as it is the case with physics and chemistry. This is due to the fact that biological and medical processes are concerned with living organisms. These processes, even under physical and chemical aspects, are too complicated in order to be fully described by a mathematical model. In addition they can contain features which in principle cannot be described mathematically. Mathematical models, however, can be used as a language by which certain aspects of biological or medical processes can be expressed.

So, for instance, in a growth model for a population of men or animals the assumption that the growth rate depends on the size of the population can be quantified with the use of differential equations and it can be checked with measured data to what extent the mathematical model gives a realistic description of the reality. In general several mathematical models can be designed in order to describe a biological or medical process and there is no unique criterion which model gives the best description. Coincidence with measured data is a necessary but by far not a sufficient condition.

This book consists of four chapters. The first is concerned with growth models for one or several populations. The focus is on conditions for stability or asymptotic stability of equilibrium states. The second chapter deals with a game-theoretic evolution model for one or two populations. It is mainly concerned with the concept of evolutionarily stable equilibria. In the third chapter four medical processes are described with the aid of optimal control theory and the fourth chapter deals with a mathematical model for the process of hemodialysis.

The book ends with an appendix in which some mathematical results are represented that are used in the text.

The authors want to thank Tino Krug, Dr. Jens Römer and Philipp Spee for their help in preparing the manuscript and especially Stefan Siegelmann for his outstanding skill.

April 2007

Werner Krabs
Stefan Pickl

Contents

List of Figures

1

Growth Models

1.1 A Growth Model for one Population

Historically the first model to describe the growth of a population (of men) was developed by Thomas Malthus in 1798. He assumes a constant birthrate γ and a deathrate δ per capita of the population and time unit and describes the change of the population size $p(t)$ within a time period Δt from t to $t + \Delta t$ by the formula

$$p(t + \Delta t) = p(t) + \lambda p(t)\Delta t \tag{1.1}$$

where $\lambda = \gamma - \delta$.

Formula (1.1) can also be written in the form

$$\frac{p(t + \Delta) - p(t)}{\Delta t} = \lambda p(t)$$

and passing to the limit $\Delta t \to 0$ leads to the differential equation

$$\frac{dp}{dt} = \lambda p(t)$$

with the unique solution

$$p(t) = p_0 e^{\lambda t} \tag{1.2}$$

if we prescribe $p(0) = p_0$.

Up to this point one could object that the function $p = p(t)$ can only assume integer values and can therefore not be represented by a differentiable function. With respect to the individuals of the population the growth (or decrease) indeed occurs in steps. In relation to the entire population these steps become smaller the larger the population becomes so that the size of the population can be approximated by a differentiable function, if the population is sufficiently large. The quality of this approximation will have to be tested experimentally from case to case. So according to [1] the development of the world population from 1700 to 1961 can be described rather precisely by an exponential law of the form (1.2), if λ is taken to be 0.02. For the period T of doubling the size of the world population during the time period from t to $t + T$ one obtains

$$p(t+T) = p_0 e^{\lambda(t+T)} = 2p_0 e^{\lambda t}$$

and hence

$$e^{\lambda T} = 2 \Leftrightarrow T = \frac{\ln 2}{\lambda} \tag{1.3}$$

For $\lambda = 0.02$ this leads to the value $T = 34.66$. This fits with the observation that $T = 35$ years.

The relation (1.3) for the doubling time T can also be confirmed in a time-discrete model. Here we have the relation

$$p(t+T) = 2p(t)$$

or for $t = 0$

$$p(T) = 2p(0)$$

from which with $p_0 = p(0)$ and $t_k = kT$ for $k = 0, 1, 2, \ldots$ we obtain

$$p(t_k) = 2^k p_0 = p_0 e^{k \ln 2} = p_0 e^{\frac{\ln 2}{T} t_k}. \tag{1.4}$$

So the continuous growth law (1.2) is in accordance with (1.4), if we put $\lambda = \frac{\ln 2}{T}$ which also fits with (1.3).

The model we treated so far in order to describe the growth of the world population starts with simple assumptions that seem to be plausible and lead to the growth law (1.2). This law can be accepted for sufficiently large populations and is also in accordance with the observation that between 1700 and 1961 the world population doubled every 35 years.

An extrapolation of this law into the future, however, leads to an astronomic growth of the world population which will be prevented by the limits of natural reserves.

So the question arises whether a growth law can be established which leads to a limited growth of the population.

In the middle of the nineteenth century the Dutch biomathematician Paul Verhulst proposed two models for a limited population growth (see [2]). In the first model he assumes the birthrate and deathrate to be linearly decreasing and increasing functions of the population size, respectively, being given in the form

$$\gamma(t) = \gamma_0 - \gamma_1 p(t) \text{ and } \delta(t) = \delta_0 + \delta_1 p(t)$$

with $\gamma_0 > \delta_0 > 0$, $\gamma_1 > 0$, $\delta_1 > 0$. As differential equation for the growth the equation

$$\frac{dp}{dt}(t) = \gamma(t) - \delta(t) = k(a - p(t)) \tag{1.5}$$

is taken where $k = \gamma_1 + \delta_1 > 0$ and $a = \frac{\gamma_0 - \delta_0}{\gamma_1 + \delta_1} > 0$.

With the initial condition $p(0) = p_0$ the solution of (1.5) reads

$$p(t) = a + (p_0 - a)e^{-kt}$$

obviously we have that

$$\lim_{t \to \infty} p(t) = a$$

which guarantees that the function $p = p(t)$is bounded. Graphically we have the following picture: Figure 1.1.

The second model is based on the differential equation

$$\frac{dp}{dt}(t) = ap(t) - bp(t)^2 = (a - bp(t))p(t) \tag{1.6}$$

with certain constants $a > 0, b > 0$. This shows that for small population sizes $p(t)$ the term $bp(t)^2$ is essentially smaller than $ap(t)$, if b is essentially smaller than a, and can be neglected so that the equation (1.6) describes exponential growth. After $p(t)$ has grown sufficiently the term $-bp(t)^2$ enters the scene and damps the exponential growth.

Of course, also in this case the growth law which can be derived from (1.6) has to stand the test with empirical data.

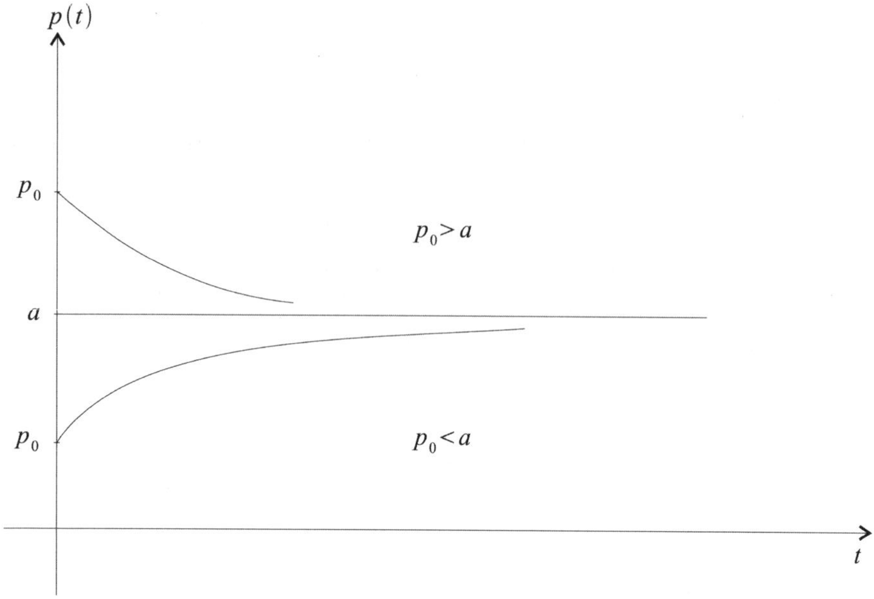

Fig. 1.1. First Verhulst Model

Under the initial condition $p(t_0) = p_0 > 0$ for some $t_0 \geq 0$ the method of separation of variables leads to

$$\begin{aligned} p(t) &= \frac{ap_0 \exp[a(t-t_0)]}{a - bp_0 + bp_0 \exp[a(t-t_0)]} \\ &= \frac{ap_0}{bp_0 + (a - bp_0)\exp[-a(t-t_0)]}. \end{aligned} \tag{1.7}$$

This representation shows that

$$\lim_{t\to\infty} p(t) = \frac{a}{b}.$$

It also shows that $p = p(t)$ is a strictly increasing and decreasing function of $t \geq t_0$, if $p_0 < \frac{a}{b}$ and $p_0 > \frac{a}{b}$, respectively. For $p_0 = \frac{a}{b}$ it follows that

$$p(t) = \frac{a}{b} = p_0$$

for all $t \geq t_0$.

A point t_S of inflection of $p = p(t)$ results from $\ddot{p}(t_S) = 0$. From the differential equation (1.6) we derive

$$\ddot{p}(t_S) = a\dot{p}(t_S) - 2bp(t_S)\dot{p}(t_S) = (a - 2bp(t_s))(a - bp(t_S))p(t_S) = 0.$$

This shows that in the case

$$p_0 > \frac{a}{b} \Rightarrow p(t) > \frac{a}{b}$$

for all $t \geq t_0$ there is no point of inflection and in the case

$$p_0 < \frac{a}{b} \Rightarrow p(t) < \frac{a}{b}$$

for all $t \geq t_0$ there is exactly one which solves the equation

$$p(t_S) = \frac{a}{2b}.$$

Graphically we have the following picture: Figure 1.2.

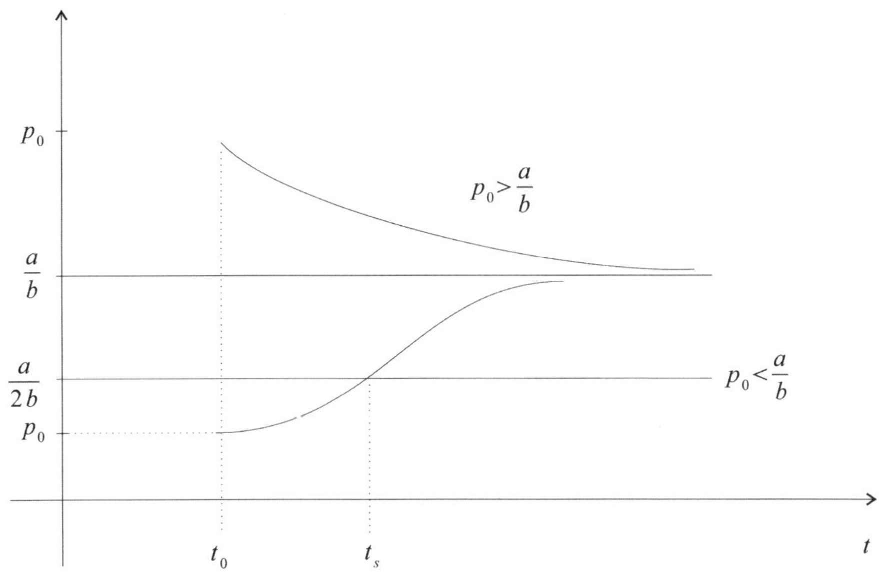

Fig. 1.2. Second Verhulst Model

If one chooses three points $t_1 < t_2 < t_3$, $t_1 \geq t_0$ with $t_2 - t_1 = t_3 - t_2$, then the parameters of the growth law (1.7) can be determined from the three values

$p(t_1), p(t_2), p(t_3)$. In the case of the population growth in the USA one obtains from the empirical values $p(1790)$, $p(1850)$ and $p(1910)$ the parameters

$$a = 0.03134 \text{ and } b = 1.5888 \cdot 10^{-10}.$$

The growth law (1.7) with these values is in considerable coincidence with empirical data (see [1]). The point t_S of inflection falls into April 1913. The equation (1.6) is not the only differential equation to describe S-shaped or logistic growth. In trying to model the growth of tumor another differential equation has been derived which is named after Gompertz. The starting point is the observation made by several researchers that the rate of growth of the tumor decreases with increasing time. This is described by a differential equation of the form

$$\frac{dp}{dt}(t) = \lambda(t)p(t) \tag{1.8}$$

where $\lambda = \lambda(t)$ is a positive strictly decreasing function. With the initial condition $p(0) = p_0$ the solution of (1.8) is given by

$$p(t) = p_0 \exp\left(\int_0^t \lambda(s)ds\right). \tag{1.9}$$

For the function $\lambda = \lambda(t)$ there are infinitely many choices. If one assume that the growth rate of the tumor decreases exponentially, i.e.

$$\lambda(t) = \lambda_0 e^{-\gamma t} \ (\lambda(0) = \lambda_0 > 0) \tag{1.10}$$

with a decreasing rate $\gamma > 0$, then one obtains

$$p(t) = p_0 \exp\left[\frac{\lambda_0}{\gamma}(1 - e^{-\gamma t})\right] \tag{1.11}$$

as growth law. Obviously it follows that

$$p_0 \le p(t) \le p_0 e^{\frac{\lambda_0}{\gamma}} \text{ and } \lim_{t \to \infty} p(t) = p_0 e^{\frac{\lambda_0}{\gamma}}.$$

The function $p = p(t)$ is strictly growing from p_0 to $p_0 e^{\frac{\lambda_0}{\gamma}}$ and has exactly one point $\hat{t} > 0$ of inflection (with $\ddot{p}(\hat{t}) = 0$), if one assumes that $\lambda_0 > \gamma$. This is then given by

$$\hat{t} = \frac{1}{\gamma} \ln \frac{\lambda_0}{\gamma}.$$

Graphically we have the following picture: Figure 1.3

By (1.11) we also have logistic growth under the assumption $\lambda_0 > \gamma$. This law is quite different from (1.7). Mathematically, a connection between these

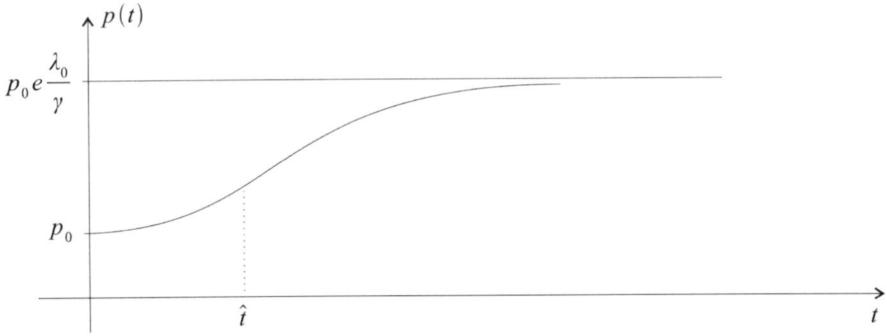

Fig. 1.3. Gompertz Model

two laws can be established. For this purpose we start with the observation that the representation (1.10), of the growth rate is equivalent to

$$\frac{d\lambda}{dt}(t) = -\gamma\lambda(t) \ , \lambda(0) = \lambda_0. \tag{1.12}$$

From (1.8) one then obtains

$$\gamma \frac{d}{dt}[\ln p(t)] = \frac{d\lambda}{dt}(t)$$

and integration of both sides gives

$$\gamma \ln\left(\frac{p(t)}{p_0}\right) = -\lambda(t) + \lambda_0$$

and hence

$$\lambda(t) = \lambda_0 - \gamma \ln\left(\frac{p(t)}{p_0}\right). \tag{1.13}$$

If one compares (1.6) with (1.8) and $\lambda(t)$ by (1.13), then one observes that both differential equations are of the form

$$\frac{dp}{dt}(t) = f(p(t))p(t) \tag{1.14}$$

with

$$f(p(t)) = a - bp(t) \tag{1.15}$$

and

$$f(p(t)) = \lambda_0 - \gamma \ln\left(\frac{p(t)}{p_0}\right), \tag{1.16}$$

respectively. In both cases $f = f(p)$, $p \geq 0$ is a strictly decreasing continuous function with $f(0) \in (0, \infty]$ which is a growth rate that decreases with increasing population size. Further we have $f(M) = 0$ for $M = \frac{a}{b}$ and $M = p_0 e^{\frac{\lambda_0}{\gamma}}$, respectively and

$$p_0 \leq p(t) \leq M \text{ for all } t \in [0, \infty) \text{ and } \lim_{t\to\infty} p(t) = M.$$

Now let us assume that the population growth is described by a differential equation of the form (1.14) where $f = f(p)$ for $p \geq 0$ is a strictly decreasing continuous function with $f(M) = 0$ for some $M > 0$. If we then prescribe, for some given $t_0 \in \mathbb{R}$, an initial population size $p_0 \in (0, M)$, i.e.

$$p(t_0) = p_0, \tag{1.17}$$

then (1.14) and (1.17) can also be written in the form

$$\int_{p_0}^{p(t)} \frac{dp}{f(p)p} = t - t_0, \; t \in \mathbb{R}. \tag{1.18}$$

Now let us define, for every $q \in [p_0, M)$,

$$F(q) = \int_{p_0}^{q} \frac{dp}{f(p)p}$$

Then $F = F(q)$ is a strictly increasing, continuously differentiable and positive function on $[p_0, M)$.

Assumption:

$$\lim_{q\to M} F(q) = \infty.$$

This assumption is satisfied for f given by (1.15) or (1.16). It implies that for every $t \geq t_0$ there exists exactly one $p(t) \in [p_0, M)$ such that

$$F(p(t)) = \int_{p_0}^{p(t)} \frac{dp}{f(p)p} = t - t_0.$$

Further $p = p(t) = F^{-1}(t - t_0)$ is differentiable and it follows

$$\frac{d}{dt}F(p(t)) = \frac{\dot{p}(t)}{f(p(t))p(t)} = 1 \text{ for all } t > t_0$$

and $p(t_0) = p_0$ as well as $\lim\limits_{t\to\infty} p(t) = M$.

Finally, $p(t) = M$ for all $t \in \mathbb{R}$ is a constant solution of (1.14).

Result: Under the above assumptions on $f = f(p)$, $p \geq 0$ there is, for every $t_0 \in \mathbb{R}$ and every $p_0 \in (0, M)$, exactly one strictly increasing solution $p = p(t)$ of (1.14) for $t > t_0$ and (1.17) with

$$p(t) \in [p_0, M) \text{ for all } t \geq t_0 \text{ and } \lim_{t\to\infty} p(t) = M.$$

If one prescribes, for some given $t_0 \in \mathbb{R}$, an initial population size $p_0 \in (M, \infty)$, i.e., (1.17) is required, then again (1.14) and (1.17) can be written in the form (1.18). Further the function

$$F(q) = \int_{p_0}^{q} \frac{dp}{f(p)p} = -\int_{q}^{p_0} \frac{dp}{f(p)p}, \quad q \in (M, p_0),$$

is continuously differentiable, positive, and strictly decreasing and it follows under the above assumption that

$$\lim_{q\to M} F(q) = \infty \text{ and } \lim_{q\to p_0} F(q) = 0.$$

Therefore, for every $t \in [t_0, \infty)$, there is exactly one $p(t) \in (M, p_0)$ with

$$F(p(t)) = t - t_0.$$

This implies (1.14) for all $t > t_0$ and $\lim\limits_{t\to\infty} p(t) = M$.

1.2 Interacting Growth of two Populations

We consider two populations of men, animals or plants whose sizes depend on time and are described by the two non-negative real valued functions $p = p(t)$ and $q = q(t)$, $t \in \mathbb{R}$. We assume the temporal development of those population sizes to be described by two differential equations of the form

$$\begin{aligned} \dot{p}(t) &= f(p(t), q(t))p(t), \\ \dot{q}(t) &= g(p(t), q(t))q(t), \quad t \in \mathbb{R}, \end{aligned} \tag{1.19}$$

with f and g being real valued functions on

$$\mathbb{R}^2_+ = \left\{(p,q) \in \mathbb{R}^2 | p \geq 0, q \geq 0\right\}$$

which can be considered as growth rates. We assume f and g to be continuously differentiable in both variables on $\mathbb{R}^2_+$. We further assume that the system

$$f(p,q) = 0 \text{ and } g(p,q) = 0 \tag{1.20}$$

has a solution $p = \hat{p} > 0$ and $q = \hat{q} > 0$. Then obviously

$$p(t) = \hat{p},\; q(t) = \hat{q} \text{ for all } t \in \mathbb{R} \tag{1.21}$$

is a solution of (1.19) which is called (for good reasons) an equilibrium state of (1.19).

In the following we are concerned with the question how the solutions of (1.19) behave in a neighbourhood of this equilibrium state. In particular we are interested in the question under which conditions an equilibrium state (1.21) with $(\hat{p}, \hat{q}) \in \overset{\circ}{\mathbb{R}}{}^2_+$ = interior of $\mathbb{R}^2_+$ is asymptotically stable, i.e.,

1. for every neighbourhood $\mathcal{U}(\hat{p}, \hat{q}) \subseteq \overset{\circ}{\mathbb{R}}{}^2_+$ of $(\hat{p}, \hat{q})$ there is a neighbourhood $\mathcal{W}(\hat{p}, \hat{q}) \subseteq \mathcal{U}(\hat{p}, \hat{q})$ such that for every $(p_0, q_0) \in \mathcal{W}(\hat{p}, \hat{q})$ the corresponding solution $(p(t),\; q(t))$ of (1.19) with $p(0) = p_0$ and $q(0) = q_0$ satisfies
$$(p(t), q(t)) \in \mathcal{U}(\hat{p}, \hat{q}) \text{ for all } t > 0 \text{ and}$$
2. there is a neighbourhood $\mathcal{U}^0(\hat{p}, \hat{q}) \subseteq \overset{\circ}{\mathbb{R}}{}^2_+$ of $(\hat{p}, \hat{q})$ such that for every $(p_0, q_0) \in \mathcal{U}^0(\hat{p}, \hat{q})$ the corresponding solution $(p(t),\; q(t))$ of (1.19) with $p(0) = p_0$ and $q(0) = q_0$ satisfies
$$\lim_{t\to\infty} (p(t), q(t)) = (\hat{p}, \hat{q}).$$

If only condition 1) is satisfied, the equilibrium state (1.21) is called stable.

It is well known (see, for instance, [4], Satz 1.12*) that (1.21) with $(\hat{p},\hat{q}) \subseteq \overset{\circ}{\mathbb{R}}{}^2_+$ is asymptotically stable, if the Jacobi matrix of

$$F(p,q) = \begin{pmatrix} f(p,q)p \\ g(p,q)q \end{pmatrix}, \quad (p,q) \in \overset{\circ}{\mathbb{R}}{}^2_+$$

in $(\hat{p},\hat{q})$ which is given by

$$J_F(\hat{p},\hat{q}) = \begin{pmatrix} F_{1p}(\hat{p},\hat{q}) & F_{1q}(\hat{p},\hat{q}) \\ F_{2p}(\hat{p},\hat{q}) & F_{2q}(\hat{p},\hat{q}) \end{pmatrix} = \begin{pmatrix} f_p(\hat{p},\hat{q})\hat{p} & f_q(\hat{p},\hat{q})\hat{p} \\ g_p(\hat{p},\hat{q})\hat{q} & g_q(\hat{p},\hat{q})\hat{q} \end{pmatrix}$$

has only eigenvalues $\lambda_1, \lambda_2 \in \mathbb{C}$ with $Re(\lambda_1) < 0$ and $Re(\lambda_2) < 0$. These eigenvalues are the solutions of the quadratic equation

$$\lambda^2 - (f_P(\hat{p},\hat{q})\hat{p} + g_q(\hat{p},\hat{q})\hat{q})\lambda + (f_p(\hat{p},\hat{q})g_q(\hat{p},\hat{q}) - f_q(\hat{p},\hat{q})g_p(\hat{p},\hat{q}))\hat{p}\hat{q} = 0$$

and are given by

$$\begin{aligned}\lambda_{1,2} = {} & \tfrac{1}{2}(f_p(\hat{p},\hat{q})\hat{p} + g_q(\hat{p},\hat{q})\hat{q}) \\ & \pm \sqrt{\tfrac{1}{4}\left(f_p(\hat{p},\hat{q})\hat{p} + g_q(\hat{p},\hat{q})\hat{q}\right)^2 - \left(f_p(\hat{p},\hat{q})g_q(\hat{p},\hat{q}) - f_q(\hat{p},\hat{q})g_p(\hat{p},\hat{q})\right)\hat{p}\hat{q}}.\end{aligned} \tag{1.22}$$

Special cases:

1. Competition: We assume that both populations fight for the same living space and that their growth rates decrease with increasing sizes of both populations. In mathematical terms this means that

$$f_p(p,q) < 0,\; f_q(p,q) < 0,\; g_p(p,q) < 0,\; g_q(p,q) < 0 \text{ for all } (p,q) \in \overset{\circ}{\mathbb{R}}{}^2_+. \tag{1.23}$$

From (1.22) it then follows that $Re(\lambda_{1,2}) < 0$, if the condition

$$f_p(\hat{p},\hat{q})g_q(\hat{p},\hat{q}) - f_q(\hat{p},\hat{q})g_p(\hat{p},\hat{q}) > 0 \tag{1.24}$$

is satisfied.

If one chooses in particular

$$\begin{aligned} f(p,q) &= a + bp + cq, \\ g(p,q) &= d + ep + fq, \end{aligned} \tag{1.25}$$

then the assumption (1.23) is equivalent to

$$b < 0,\; c < 0,\; e < 0,\; f < 0 \tag{1.26}$$

and the condition (1.24) reads

$$bf - ce > 0. \tag{1.27}$$

This implies that the system

$$\begin{aligned} f(p,q) &= a + bp + cq = 0, \\ g(p,q) &= d + ep + fq = 0 \end{aligned} \tag{1.28}$$

has exactly one solution $p = \hat{p},\ q = \hat{q}$ which is given by

$$\hat{p} = \frac{cd - af}{bf - ce},\ \hat{q} = \frac{ae - bd}{bf - ce}. \tag{1.29}$$

Further it follows that $\hat{p} > 0$ and $\hat{q} > 0$, if and only if

$$cd - af > 0 \text{ and } ae - bd > 0. \tag{1.30}$$

Thus we have the following statement: The system (1.19) with f and g by (1.25) and under the assumption (1.26) has exactly one equilibrium state (1.21) with $\hat{p} > 0$ and $\hat{q} > 0$ given by (1.29), if the conditions (1.27) and (1.30) are satisfied.

2. Predator-prey behavior: We assume that the p-population serves as prey to the q-population. Then its growth rate decreases as its own size and the size of the predator-(q)-population increases. In mathematical terms this means that

$$f_p(p,q) < 0 \text{ and } f_q(p,q) < 0 \text{ for all } (p,q) \in \overset{\circ}{\mathbb{R}}{}_+^2. \tag{1.31}$$

The growth rate of the q-population however only decreases, if its own size increases and increases, if the size of die p-population increases. Mathematically this leads to

$$g_p(p,q) > 0 \text{ and } g_q(p,q) < 0 \text{ for all } (p,q) \in \overset{\circ}{\mathbb{R}}{}_+^2. \tag{1.32}$$

We again assume that there exists an equilibrium state (1.21) with $(\hat{p}, \hat{q}) \in \overset{\circ}{\mathbb{R}}{}_+^2$ being a solution of the system

$$f(\hat{p}, \hat{q}) = 0 \text{ and } g(\hat{p}, \hat{q}) = 0. \tag{1.33}$$

From (1.22) it then follows that $Re(\lambda_{1,2}) < 0$ without any further condition to be satisfied.

Therefore the equilibrium state (1.21) with $(\hat{p}, \hat{q}) \in \overset{\circ}{\mathbb{R}}{}_+^2$ being a solution of (1.33) is asymptotically stable.

If we assume the growth rates f and g in (1.19) to be of the form (1.25) then the conditions (1.31) and (1.32) are equivalent to

$$b < 0, \ c < 0, \ e > 0, \ f < 0. \tag{1.34}$$

Thus condition (1.27) is satisfied. If in addition the conditions (1.30), (1.34) are satisfied then the system (1.19) with f and g by (1.25) and under the assumption (1.34) has exactly one equilibrium state (1.21) with $\hat{p} > 0$ and $\hat{q} > 0$ given by (1.29).

The same result can be obtained by Lyapunov's method. For that purpose we consider the mapping

$$u = \ln \frac{p}{\hat{p}}, \ v = \ln \frac{q}{\hat{q}} \ (\hat{p} > 0, \hat{q} > 0) \text{ of } \overset{\circ}{\mathbb{R}}{}_+^2 \text{ on } \mathbb{R}^2.$$

If $p = p(t), q = q(t), t \in \mathbb{R}$ is a solution of the system

$$\begin{aligned} \dot{p}(t) &= (a + bp(t) + cq(t))p(t), \\ \dot{q}(t) &= (d + ep(t) + fq(t))q(t), \ t \in \mathbb{R} \end{aligned} \tag{1.35}$$

with $p(t) > 0, q(t) > 0$ for all $t \in \mathbb{R}$ and if for $\hat{p} > 0$ and $\hat{q} > 0$ we have

$$\begin{aligned} a + b\hat{p} + c\hat{q} &= 0, \\ d + e\hat{p} + f\hat{q} &= 0, \end{aligned} \tag{1.36}$$

then the system (1.35) transforms itself to into the system

$$\begin{aligned} \dot{u}(t) &= b\hat{p}(e^{u(t)} - 1) + c\hat{q}(e^{v(t)} - 1), \\ \dot{v}(t) &= e\hat{p}(e^{u(t)} - 1) + f\hat{q}(e^{v(t)} - 1), \ t \in \mathbb{R}. \end{aligned} \tag{1.37}$$

This system has

$$u(t) = v(t) = 0 \text{ for all } t \in \mathbb{R} \tag{1.38}$$

as unique equilibrium state and this is asymptotically stable, if and only if the equilibrium state

$$p(t) = \hat{p}, \ q(t) = \hat{q} \text{ for all } t \in \mathbb{R} \tag{1.39}$$

of (1.35) is asymptotically stable.

In order to show the asymptotical stability of the equilibrium state (1.38) of the system (1.37) we define a Lyapunov function $V : \mathbb{R}^2 \to \mathbb{R}$ by

$$V(u,v) = e\hat{p}(e^u - u) - c\hat{q}(e^v - v) - e\hat{p} + c\hat{q}, \ u, v \in \mathbb{R}.$$

Then it follows that

$$V(0,0) = 0 \text{ and } V(u,v) > 0 \text{ for all } (u,v) \neq (0,0)$$

Furthermore it follows with

$$F(u,v) = \begin{pmatrix} b\hat{p}(e^u - 1) + c\hat{q}(e^v - 1) \\ e\hat{p}(e^u - 1) + f\hat{q}(e^v - 1) \end{pmatrix}, \ u, v \in \mathbb{R}. \tag{1.40}$$

that

$$\begin{aligned} \dot{V}(u,v) &= grad\ V(u,v)^T F(u,v) \\ &= be\hat{p}^2(e^u - 1)^2 - cf\hat{q}^2(e^v - 1)^2 < 0 \end{aligned}$$

for all $(u,v) \neq (0,0)$ and $\dot{V}(0,0) = 0$.

By Satz 1.12 in [4] we therefore infer that the equilibrium state (1.38) of the system (1.37) is asymptotically stable and hence the equilibrium state (1.39) of the system (1.35) as well.

If one replaces the conditions (1.34) by

$$c < 0, \ e > 0 \text{ and } b = f = 0 \tag{1.41}$$

then one obtains the classical Volterra-Lotka-model in which it is assumed that $a > 0$ and $d < 0$. This implies that in the absence of predators the prey grows exponentially and in the absence of prey the predators decrease exponentially. The conditions (1.27) and (1.30) are satisfied and the only equilibrium state of the system (1.35) is given by (1.39) with:

$$\hat{p} = -\frac{d}{e} \text{ and } \hat{q} = -\frac{a}{c}$$

Further it also follows that

$$V(0,0) = 0 \text{ and } V(u,v) > 0 \text{ for all } (u,v) \neq (0,0)$$

However, it follows that

$$\dot{V}(u,v) = 0 \text{ for all } (u,v) \in \mathbb{R}^2.$$

Hence, by Satz 1.12 in [4] it only follows that the equilibrium state (1.38) of the system (1.37) is stable and in turn also the equilibrium state (1.39) of the system (1.35).

It is unrealistic to assume that the prey population grows exponentially in the absence of predators. So it is reasonable to assume instead of (1.41) that

$$b < 0, \ c < 0, \ e > 0, \ f = 0$$

which ensures limited growth of the prey population in the absence of predators. But then we have

$$\dot{V}(0, v) = 0 \text{ for all } v \in \mathbb{R}$$

so that asymptotical stability of the equilibrium state (1.38) of the system (1.37) cannot be inferred by the above choice of a Lyapunov function.

However, the eigenvalues of the Jacobi matrix $J_F(0, 0)$ of F given by (1.40) read

$$\lambda_{1,2} = \frac{1}{2} b\hat{p} \pm \sqrt{\frac{1}{4}(b\hat{p})^2 + ec\hat{p}\hat{q}}$$

so that $Re(\lambda_{1,2}) < 0$ which implies that the equilibrium state (1.38) of the system (1.37) is asymptotically stable.

1.3 Interacting Growth of $n \geq 2$ Populations

We consider $n \geq 2$ populations whose sizes depend on time and are described by non-negative real valued functions $p_i = p_i(t), \ t \in \mathbb{R}, \ i = 1, \ldots, n$. We assume the temporal development of these population sizes to be described by n differential equations of the form

$$\dot{p}_i(t) = f_i(p(t))p_i(t), \ t \in \mathbb{R}, \text{ for } i = 1, \ldots, n \tag{1.42}$$

and $p(t) = (p_1(t), \ldots, p_n(t))$ where $f_i = f_i(p) = f_i(p_1, \ldots, p_n), \ i = 1, \ldots n$, are real valued functions on

$$\mathbb{R}^n_+ = \{p \in \mathbb{R}^n \mid p_i \geq 0 \text{ for } i = 1, \ldots, n\}$$

which can be considered as growth rates.

We assume the $f_i, \ i = 1, \ldots, n$, to be continuously differentiable on $\mathbb{R}^n_+$ with respect to all variables.

We further assume that the system

$$f_i(p) = 0 \text{ for } i = 1, \ldots, n \tag{1.43}$$

has a solution $p = \hat{p} \in \overset{\circ}{\mathbb{R}}{}^n_+ = \{p \in \mathbb{R}^n | p_i > 0 \text{ for } i = 1, \ldots, n\}$. Then obviously

$$p(t) = \hat{p}, \; t \in \mathbb{R}, \tag{1.44}$$

is a solution of (1.42) which is called an equilibrium state of (1.42). In the following we are again concerned with the question how the solutions of (1.42) behave in a neighbourhood of this equilibrium state. In particular we are interested in the question under which conditions an equilibrium state (1.44) with $\hat{p} \in \overset{\circ}{\mathbb{R}}{}^n_+$ is asymptotically stable, i.e.

1. for every neighbourhood $\mathcal{U}(\hat{p}) \subseteq \overset{\circ}{\mathbb{R}}{}^n_+$ of $\hat{p}$ there is a neighbourhood $W(\hat{p}) \subseteq \mathcal{U}(\hat{p})$ so that for every $p_0 \in W(\hat{p})$ the corresponding solution $p = p(t)$ of (1.42) with $p(0) = p_0$ satisfies

$$p(t) \in \mathcal{U}(\hat{p}), \text{ for all } t > 0 \text{ and}$$

2. there is a neighbourhood $\mathcal{U}^0(\hat{p}) \subseteq \overset{\circ}{\mathbb{R}}{}^n_+$ of $\hat{p}$ such that for every $p_0 \in \mathcal{U}^0(\hat{p})$ the corresponding solution $p = p(t)$ of (1.42) satisfies

$$\lim_{t \to \infty} p(t) = \hat{p}.$$

If only condition 1) is satisfied, the equilibrium state (1.44) with $\hat{p} \in \overset{\circ}{\mathbb{R}}{}^n_+$ is called stable. It is well known (see, for instance [4], Satz 1.12*) that the equilibrium state (1.44) with $\hat{p} \in \overset{\circ}{\mathbb{R}}{}^n_+$ is asymptotically stable, if the Jacobi matrix of

$$F(p) = \begin{pmatrix} f_1(p)p_1 \\ \vdots \\ f_n(p)p_n \end{pmatrix}, \quad p \in \overset{\circ}{\mathbb{R}}{}^n_+,$$

in $\hat{p}$ which is given by

$$J_F(\hat{p}) = \begin{pmatrix} f_{1p_1}(\hat{p})\hat{p}_1 & \cdots & f_{1p_n}(\hat{p})\hat{p}_1 \\ \vdots & & \vdots \\ f_{np_1}(\hat{p})\hat{p}_n & \cdots & f_{np_n}(\hat{p})\hat{p}_n \end{pmatrix}$$

has only eigenvalues with negative real parts.

Let $n = 3$. If we then define, for $F(p) = (F_1(p), F_2(p), F_3(p))$,

$$\begin{aligned} a_1 &= -F_{1p_1}(\hat{p}) - F_{2p_2}(\hat{p}) - F_{3p_3}(\hat{p}), \\ a_2 &= F_{1p_1}(\hat{p})F_{2p_1}(\hat{p}) - F_{1p_2}(\hat{p})F_{2p_1}(\hat{p}) + F_{1p_1}(\hat{p})F_{3p_3}(\hat{p}) - F_{1p_3}(\hat{p})F_{3p_1}(\hat{p}) \\ &\quad + F_{2p_2}(\hat{p})F_{3p_3}(\hat{p}) - F_{2p_3}(\hat{p})F_{3p_2}(\hat{p}), \\ a_3 &= -\det J_F(\hat{p}) \end{aligned}$$

by a Theorem of Hurwitz (see [3]), the Jacobi matrix $J_F(\hat{p})$ has only eigenvalues with negative real parts, if and only if

$$a_1 > 0,\ a_1 a_2 - a_3 > 0,\ a_3 > 0. \tag{1.45}$$

Now let us assume in particular that in all three populations P_1, P_2, P_3 the growth rate $f_i(p)$, $i = 1, 2, 3$, decreases with growing population size P_i which implies

$$F_{p_i}(\hat{p}) = f_{ip_i}(\hat{p})\hat{p}_i < 0 \text{ for } i = 1, 2, 3. \tag{1.46}$$

We further assume that the populations P_2 and P_3 have the population P_1 as prey and are neutral to each other. Mathematically this leads to the following conditions

$$\begin{aligned} &F_{1p_2}(\hat{p}) < 0,\ F_{1p_3}(\hat{p}) < 0, \\ &F_{2p_1}(\hat{p}) > 0,\ F_{3p_1}(\hat{p}) > 0, \\ &F_{2p_3}(\hat{p}) = 0,\ F_{3p_2}(\hat{p}) = 0. \end{aligned} \tag{1.47}$$

From (1.46) it follows immediately that $a_1 > 0$ and (1.46), (1.47) imply that $a_1 a_2 - a_3 > 0$ and $a_3 > 0$. Hence, under the assumptions (1.46), (1.47) the equilibrium state (1.44) of the system (1.42) is asymptotically stable. *A special case:* We assume that the growth rates f_i in (1.42) are of the form

$$f_i(p) = c_i + \sum_{j=1}^{n} c_{ij} p_j,\ p \in \mathbb{R}_+^n,\ i = 1, \ldots, n. \tag{1.48}$$

The system (1.43) is then linear and reads for $p = \hat{p} \in \overset{\circ}{\mathbb{R}}{}_+^n$

$$\sum_{j=1}^{n} c_{ij}\hat{p}_j = -c_i,\ i = 1, \ldots, n. \tag{1.49}$$

The Jacobi matrix $J_F(\hat{p})$ is given by

$$J_F(\hat{p}) = \begin{pmatrix} c_{11}\hat{p}_1 & \cdots & c_{1n}\hat{p}_1 \\ \vdots & & \vdots \\ c_{n1}\hat{p}_n & \cdots & c_{nn}\hat{p}_n \end{pmatrix}.$$

We assume that in all populations $P_i,\ i = 1,\ldots,n$ the growth rate $f_i(p)$ decreases with growing population size $p_i,\ i = 1,\ldots,n$, which implies

$$c_{ii} < 0 \text{ for } i = 1,\ldots,n. \tag{1.50}$$

We further assume that the populations live in mutual predator-prey relations or are neutral to each other. This leads to the conditions ($i \neq j$):

$$\begin{aligned} c_{ij} &> 0, \text{ if } P_i = \text{predator and } P_j = \text{prey}, \\ c_{ij} &< 0, \text{ if } P_i = \text{prey and } P_j = \text{predator}, \\ c_{ij} = c_{ji} &= 0, \text{ if } P_i \text{ and } P_j \text{ are neutral to each other.} \end{aligned}$$

We again consider the case $n = 3$ and again assume that the populations P_2 and P_3 have the population P_1 as prey and are neutral to each other. Then we have (1.50) for $n = 3$ and

$$c_{12} < 0,\ c_{13} < 0,\ c_{21} > 0,\ c_{31} > 0,\ c_{23} = c_{32} = 0. \tag{1.51}$$

The system (1.49) reads

$$\begin{aligned} c_{11}\hat{p}_1 + c_{12}\hat{p}_2 + c_{13}\hat{p}_3 &= -c_1, \\ c_{21}\hat{p}_1 + c_{22}\hat{p}_2 \qquad\qquad &= -c_2, \\ c_{31}\hat{p}_1 \qquad\qquad + c_{33}\hat{p}_3 &= -c_3. \end{aligned} \tag{1.52}$$

It has the unique solutions

$$\begin{aligned} \hat{p}_1 &= \tfrac{1}{\Delta}(-c_{22}c_{33}c_1 + c_{12}c_{33}c_2 + c_{13}c_{22}c_3) \\ \hat{p}_2 &= \tfrac{1}{\Delta}(-c_{11}c_{33}c_2 + c_{21}c_{33}c_1 + c_{13}(-c_{21}c_3 + c_{31}c_2)) \\ \hat{p}_3 &= \tfrac{1}{\Delta}(-c_{11}c_{22}c_3 - c_{12}(-c_{21}c_3 + c_{31}c_2) + c_{22}c_{21}c_1) \end{aligned} \tag{1.53}$$

where

$$\Delta = c_{11}c_{22}c_{33} - c_{12}c_{21}c_{33} - c_{13}c_{31}c_{22} < 0. \tag{1.54}$$

If we additionally assume that

$$c_1 > 0,\ c_2 < 0,\ c_3 < 0 \tag{1.55}$$

then $\hat{p}_1 > 0$ is implied and from the further conditions

$$c_{31}c_2 - c_{21}c_3 = 0, c_{21}c_1 - c_{11}c_2 > 0, c_{31}c_1 - c_{11}c_3 > 0 \tag{1.56}$$

it follows that $\hat{p}_2 > 0$ and $\hat{p}_3 > 0$. This leads to the following

Result: Under the conditions (1.50) for $n = 3$, (1.51), (1.55) and (1.56) the system (1.52) has exactly one solution $\hat{p} \in \overset{\circ}{\mathbb{R}}{}^3_+$ which is given by (1.53), (1.54) and

$$p_i(t) = \hat{p}_i, \ i = 1, 2, 3, \ t \in \mathbb{R}$$

is an asymptotically stable equilibrium state of the system

$$\dot{p}_i(t) = \left(c_i + \sum_{j=1}^{3} c_{ij} p_j(t) \right) p_i(t), \ i = 1, 2, 3, \ t \in \mathbb{R}.$$

If $n > 3$, the method that has been applied so far in order to show the asymptotical stability of equilibrium states becomes rather complicated. Therefore we will again apply Lyapunov's method. For that purpose we assume that for every $p_0 \in \overset{\circ}{\mathbb{R}}{}^n_+$ there is exactly one solution $p = p(t)$, $t \in \mathbb{R}$, of the system (1.42) with $p(0) = p_0$ and $p_i(t) > 0$ for all $t \in \mathbb{R}$ and $i = 1, \ldots, n$,. For each such we then define functions $u_i : \mathbb{R} \to \mathbb{R}^n$, $i = 1, \ldots, n$, by

$$u_i(t) = \ln \left(\frac{p_i(t)}{\hat{p}_i} \right) \Leftrightarrow p_i(t) = \hat{p}_i e^{u_i(t)}, \tag{1.57}$$

where

$$f_i(\hat{p}) = 0 \text{ for } i = 1, \ldots, n.$$

Then it follows that

$$\dot{u}_i(t) = \frac{\dot{p}_i(t)}{p_i(t)} = f_i(p(t)) = f_i\left(\hat{p}_1 e^{u_1(t)}, \ldots, \hat{p}_n e^{u_n(t)}\right), \ t \in \mathbb{R}, \ \text{ for } i = 1, \ldots, n. \tag{1.58}$$

If we define

$$g_i(u) = g_i(u_1, \ldots, u_n) = f_i(\hat{p}_1 e^{u_1}, \ldots, \hat{p}_n e^{u_n}), \ u \in \mathbb{R}^n, \ \text{ for } i = 1, \ldots, n,$$

then the system (1.58) can be written in the form

$$\dot{u}_i(t) = g_i(u(t)), \ t \in \mathbb{R}, \ i = 1, \ldots, n, \tag{1.59}$$

which turns out to be equivalent to the system (1.42) via the transformation (1.57). Further we have $g_i(\Theta_n) = f_i(\hat{p})$ for $i = 1, \ldots, n$ and hence

$$f_i(\hat{p}) = 0 \iff g_i(\Theta_n) = 0.$$

Finally it follows that

$$J_g(\Theta_n) = J_F(\hat{p})^T$$

and

$$u_i(t) = 0 \text{ for } i = 1, \ldots, n$$

is an asymptotically stable equilibrium state of the system (1.59), if and only if (1.44) is an asymptotically stable equilibrium state of the system (1.42).

If the $f_i,\ i = 1, \ldots, n$ are of the form (1.48), the system (1.59) reads

$$\dot{u}_i(t) = \sum_{j=1}^{n} c_{ij}\hat{p}_j(e^{u_j(t)} - 1),\ i = 1, \ldots, n, t \in \mathbb{R}, \tag{1.60}$$

and is equivalent to the system

$$\dot{p}_i(t) = \left(c_i + \sum_{j=1}^{n} c_{ij}p_j(t)\right)p_i(t),\ i = 1, \ldots, n,\ t \in \mathbb{R}, \tag{1.61}$$

if $\hat{p} \in \overset{\circ}{\mathbb{R}}{}^n_+$ is a solution of the system (1.49).

Now we define a Lyapunov function $V : \mathbb{R}^n \to \mathbb{R}$ by

$$V(u_1, \ldots, u_n) = \sum_{j=1}^{n} (-c_{ij})\hat{p}_i(e^{u_i} - u_i - 1),\ u_i \in \mathbb{R} \text{ for } i = 1, \ldots, n.$$

Then it follows that

$$V(u_1, \ldots, u_n) \geq 0 \text{ for all } (u_1, \ldots, u_n) \in \mathbb{R}^n$$

and

$$V(u_1, \ldots, u_n) = 0 \iff (u_1, \ldots, u_n) = (0, \ldots, 0).$$

Further we imply for $g(u_1, \ldots, u_n) = (g_1(u_1, \ldots, u_n), \ldots, g_n(u_1, \ldots, u_n))$ and

$$g_i(u_1, \ldots, u_n) = \sum_{j=1}^{n} c_{ij}\hat{p}_j(e^{u_j} - 1) \text{ for } i = 1, \ldots, n$$

that

$$\begin{aligned}
&\text{grad } V(u_1, \ldots, u_n)^T g(u_1, \ldots, u_n) \\
&= \sum_{i=1}^{n} (-c_{ij}^2)\hat{p}_i^2(e^{u_i} - 1)^2 + \sum_{\substack{i,j=1 \\ i<j}}^{n} \left[(-c_{ii})c_{ij} + (-c_{jj}c_{ji})\right]\hat{p}_i\hat{p}_j(e^{u_i} - 1)(e^{u_j} - 1)
\end{aligned}$$

Assumption:

$$c_{ii}c_{ij} + c_{jj}c_{ji} = 0 \text{ for all } i, j = 1, \ldots, n \text{ with } i < j. \tag{1.62}$$

From this assumption we infer

$$\operatorname{grad} V(u_1, \ldots, u_n)^T g(u_1, \ldots, u_n) \leq 0 \text{ for all } (u_1, \ldots, u_n) \in \mathbb{R}^n$$

and

$$\operatorname{grad} V(u_1, \ldots, u_n)^T g(u_1, \ldots, u_n) = 0 \iff (u_1, \ldots, u_n) = (0, \ldots, 0).$$

By Satz 1.12 in [4] it therefore follows that $u(t) = \Theta_n$ for $t \in \mathbb{R}$ is an asymptotically stable equilibrium state of the system (1.60) and hence $p(t) = \hat{p},\ t \in \mathbb{R}$ with $\hat{p} \in \overset{\circ}{\mathbb{R}}{}^n_+$ being a solution of the system (1.49) an asymptotically stable equilibrium state of the system (1.61). For $n = 3$ under the conditions (1.50), (1.51) the conditions (1.62) read

$$\begin{aligned} c_{11}c_{12} + c_{22}c_{21} &= 0, \\ c_{11}c_{13} + c_{33}c_{31} &= 0. \end{aligned} \tag{1.63}$$

Under the assumptions (1.55), (1.56) the unique positive solutions of the system (1.52) are then given by

$$\begin{aligned} \hat{p}_1 &= -\frac{c_{11}c_1 + c_{21}c_2 + c_{31}c_3}{c_{11}^2 + c_{21}^2 + c_{31}^2}, \\ \hat{p}_2 &= \frac{1}{c_{22}} \frac{c_{11}(c_{21}c_1 - c_{11}c_2)}{c_{11}^2 + c_{21}^2 + c_{31}^2}, \\ \hat{p}_3 &= \frac{1}{c_{33}} \frac{c_{11}(c_{31}c_1 - c_{11}c_3)}{c_{11}^2 + c_{21}^2 + c_{31}^2}. \end{aligned}$$

In the case $n = 2$ under the conditions $c_{12} < 0, c_{21} > 0$ the conditions (1.62) are equivalent with

$$c_{11}c_{12} + c_{22}c_{21} = 0.$$

In Section 1.2 we have seen that this condition can be dispensed with, if as Lyapunov function $V : \mathbb{R}^2 \to \mathbb{R}$ the function

$$V(u_1, u_2) = c_{21}\hat{p}_1(e^{u_1} - u_1 - 1) - c_{12}\hat{p}_2(e^{u_2} - u_2 - 1),\ u_1, u_2 \in \mathbb{R},$$

is chosen.

A generalization to $n \geq 2$ leads to the Lyapunov function $V : \mathbb{R}^n \to \mathbb{R}$ which is given by

$$V(u_1, \ldots, u_n) = \sum_{j=1}^{n} |c_{n+1-i\,i}|\hat{p}_i(e^{u_i} - u_i - 1) \text{ for } (u_1, \ldots, u_n) \in \mathbb{R}^n.$$

If we assume

$$c_{n+1-i\,i} \neq 0 \text{ for all } i = 1, \ldots, n,$$

then it follows that

$$V(u_1, \ldots, u_n) \geq 0 \text{ for all } (u_1, \ldots, u_n) \in \mathbb{R}^n$$

and

$$V(u_1, \ldots, u_n) = 0 \iff (u_1, \ldots, u_n) = (0, \ldots, 0).$$

Further it follows for

$$g(u_1, \ldots, u_n) = (g_1(u_1, \ldots, u_n), \ldots, g_n(u_1, \ldots, u_n))$$

and

$$g_i(u_1, \ldots, u_n) = \sum_{j=1}^{n} c_{ij}\hat{p}_j(e^{u_j} - 1) \text{ for } i = 1, \ldots, n$$

that

$$\text{grad } V(u_1, \ldots, u_n)^T g_i(u_1, \ldots, u_n) = \sum_{i=1}^{n} |c_{n+1-i\,i}|c_{ii}\,\hat{p}_i^2(e^{u_i} - 1)^2$$
$$+ \sum_{i=1}^{n} \sum_{j>i} (|c_{n+1-i\,i}|c_{ij} + |c_{n+1-j\,j}|c_{ji})\hat{p}_i\hat{p}_j(e^{u_i} - 1)(e^{u_j} - 1).$$

Assumption

$$|c_{n+1-i\,i}|c_{ij} + |c_{n+1-j\,j}|c_{ji} = 0 \text{ for all } i, j = 1, \ldots, n \text{ with } i < j. \tag{1.64}$$

This assumption implies

$$\text{grad } V(u_1, \ldots, u_n)^T g(u_1, \ldots, u_n) \leq 0 \text{ for all } (u_1, \ldots, u_n) \in \mathbb{R}^n$$

and

$$\text{grad } V(u_1, \ldots, u_n)^T g(u_1, \ldots, u_n) = 0 \iff (u_1, \ldots, u_n) = (0, \ldots, 0).$$

Again by Satz 1.12 in [4] $u(t) = \Theta_n$ for $t \in \mathbb{R}$ is an asymptotically stable equilibrium state of the system (1.60) and in turn $p(t) = \hat{p},\ t \in \mathbb{R}$, with $\hat{p} \in \overset{\circ}{\mathbb{R}}{}_+^n$

being a solution of the system (1.49) an asymptotically stable equilibrium state of the system (1.61).

For $n = 2$ and $c_{12} < 0,\ c_{21} > 0$ the assumption (1.64) is satisfied for it is

$$|c_{21}|c_{12} + |c_{12}|c_{21} = c_{21}c_{12} - c_{12}c_{21} = 0\ (i = 1, j = 2).$$

For $n = 3$ the assumption (1.64) is equivalent with

$$\begin{aligned} |c_{31}|c_{12} + |c_{22}|c_{21} &= 0,\ \ (i = 1, j = 2), \\ |c_{31}|c_{13} + |c_{13}|c_{31} &= 0,\ \ (i = 1, j = 3), \\ |c_{22}|c_{23} + |c_{13}|c_{32} &= 0,\ \ (i = 2, j = 3). \end{aligned}$$

Under the assumptions (1.50) and (1.51) the assumption (1.64) turns out to be equivalent to

$$c_{31}c_{12} - c_{22}c_{21} = 0.$$

In [5], [6] we have given further definitions of Lyapunov functions by which stability or asymptotical stability of equilibrium states can be shown.

1.4 Discretization of the Time-Continuous Model

1.4.1 The n-Population Model

We start with the model (1.42) for the description of the temporal development of the population sizes in an interacting growth model of $n \geq 2$ population where we assume that the functions $f_i,\ i = 1, \ldots, n$, are continuously differentiable with respect to all variables on $\mathbb{R}^n$. We discretize this model by introducing a time step size $h > 0$ and replacing in (1.42) the derivates $\dot{p}_i$ by difference quotients

$$\frac{p_i(t+h) - p_i(t)}{h},\ \ i = 1, \ldots, n$$

Thereby we obtain from (1.42) the following system of difference equations

$$p_i(t+h) = (1 + hf_i(p(t))p_i(t),\ \ t \in \mathbb{R},\ \ \text{for } i = 1, \ldots, n. \tag{1.65}$$

If we define a vector function $g^h : \mathbb{R}^n \to \mathbb{R}^n$ by

$$g^h(p) = (1 + hf_i(p))p_i,\ \ i = 1, \ldots, n,\ \ p \in \mathbb{R}^n,$$

then we obtain a continuous function and with the definition

$$\Pi_h(p,k) = (g^h)^k(p) = \underbrace{g^h \cdot g^h \cdots g^h}_{\text{k times}}(p) \text{ for } p \in \mathbb{R}^n, k \in \mathbb{N},\ \Pi_h(p,0) = p,\ p \in \mathbb{R}^n,$$

we obtain a time-discrete dynamical system (see [4]) which is called a discretization of (1.42) (with stepsize h). A point $\hat{p} \in \overset{\circ}{\mathbb{R}}{}^n_+$ is a solution of $f(\hat{p}) = \Theta_n$ (i.e., $p(t) = \hat{p},\ t \in \mathbb{R}$, is an equilibrium state of the system (1.42)), if and only if $\hat{p}$ is a fixed point of g^h, i.e., a solution of the equation $g^h(\hat{p}) = \hat{p}$.

The Jacobi matrix of g^h in $\hat{p} \in \overset{\circ}{\mathbb{R}}{}^n_+$ reads

$$J_{gh}(\hat{p}) = (\delta_{ij} + hf_{ip_j}(\hat{p})\hat{p}_i)_{i,j=1,\dots,n}$$

where

$$\delta_{ij} = \begin{cases} 0, & \text{for } i \neq j \\ 1, & \text{for } i = j \end{cases}.$$

In matrix formulation we have

$$J_{gh}(\hat{p}) = I_n + hJ_F(\hat{p}), \tag{1.66}$$

$I_n = n \times n$-unit matrix.

Therefore $\lambda \in \mathbb{C}$ is an eigenvalue of $J_F(\hat{p})$, if and only if $1+h\lambda$ is an eigenvalue of $J_{gh}(\hat{p})$. Further we obtain

$$\begin{aligned} |1 + h\lambda|^2 &= (1 + Re(\lambda)h)^2 + (Im(\lambda)h)^2 \\ &= 1 + 2Re(\lambda)h + h^2|\lambda|^2 \\ &= 1 + h(2Re(\lambda) + h|\lambda|^2). \end{aligned}$$

From this equation it follows that $|1+h\lambda| < 1$, if and only if $2Re(\lambda)+h|\lambda|^2 < 0$ which is equivalent with

$$Re(\lambda) < 0 \text{ and } h < \frac{-2Re(\lambda)}{|\lambda|^2}. \tag{1.67}$$

By Satz 1.19 in [4] $\hat{p} \in \overset{\circ}{\mathbb{R}}{}^n_+$ with $g^h(\hat{p}) = \hat{p}$ is an attractor with respect to g^h, i.e., there is an open neighbourhood $\mathcal{U}(\hat{p}) \subseteq \overset{\circ}{\mathbb{R}}{}^n_+$ of $\hat{p}$ with

$$\lim_{k\to\infty} \|(g^h)^k(p) - \hat{p}\|_2 = 0 \text{ for all } p \in \mathcal{U}(\hat{p}),$$

if all the eigenvalues of g^h are smaller than 1 in modulus or, equivalently, if for every eigenvalue λ of $J_F(\hat{p})$ the condition (1.67) is satisfied.

Summarizing we obtain the following

Conclusion: If, for some $\hat{p} \in \overset{\circ}{\mathbb{R}}{}^n_+$, the equation $f(\hat{p}) = \Theta_n$ is satisfied and if all the eigenvalues of $J_F(\hat{p})$ have a negative real part (which implies that the equilibrium state (1.44) of the system (1.42) is asymptotically stable), then $J_{gh}(\hat{p})$ has only eigenvalues which are smaller than 1 in modulus (which implies that $\hat{p}$ is an attractor with respect to g^h), if the step size $h > 0$ is sufficiently small.

One can even prove that under the assumption of the conclusion the fixed point $\hat{p}$ of g^h is stable. For that purpose we define a Lyapunov function $V : \mathbb{R}^n \to \mathbb{R}$ with respect to g^h by

$$V(p) = \|p - \hat{p}\|_2, \;\; p \in \mathbb{R}^n.$$

Then V is continuous and

$$V(p) \geq 0 \text{ for all } p \neq \hat{p}$$

and

$$V(p) = 0 \iff p = \hat{p}.$$

We assume that the Jacobi matrix $J_{gh}(\hat{p})$ has only eigenvalues which are smaller than 1 in modulus. Then it follows from the proof of Satz 1.19 in [4] the existence of an open neighbourhood $\mathcal{U}(\hat{p}) \subseteq \overset{\circ}{\mathbb{R}}{}^n_+$ and a constant $\gamma \in (0, 1)$ with

$$V(g^h(p)) = \|g^h(p) - \hat{p}\|_2 \leq \gamma\|p - \hat{p}\|_2 < \|p - \hat{p}\|_2 = V(p),$$
$$\text{for all } p \in \mathcal{U}(\hat{p}) \text{ with } p \neq \hat{p}.$$

By Satz 5.8 in [7] (see also Section A.3) it follows that $\hat{p} \in \overset{\circ}{\mathbb{R}}{}^n_+$ (with $g^h(\hat{p}) = \hat{p}$) is an asymptotically stable fixed point of g^h, i.e., an attractor and stable with respect to g^h which means that for every open neighbourhood $\mathcal{U}(\hat{p})$ of $\hat{p}$ there exists an open neighbourhood $\mathcal{W}(\hat{p}) \subseteq \mathcal{U}(\hat{p})$ of $\hat{p}$ with

$$(g^h)^k(p) \in \mathcal{U}(\hat{p}) \text{ for all } p \in \mathcal{W}(\hat{p}) \text{ and all } k \in \mathbb{N}_0.$$

An Example: We consider the case $n = 2$ and the system

$$\begin{aligned}\dot{p}_1(t) &= (c_1 + c_{11}p_1(t) + c_{12}p_2(t))p_1(t),\\ \dot{p}_2(t) &= (c_2 + c_{21}p_1(t) + c_{22}p_2(t))p_2(t), \;\; t \in \mathbb{R},\end{aligned}$$

where

$$c_1 > 0,\ c_2 < 0,\ c_{11} < 0,\ c_{12} < 0,\ c_{22} < 0,\ c_{21} > 0$$

(see (1.19) with f and g according to (1.25)). The system $f(\hat{p}) = \Theta_2$ then has the unique solutions

$$\hat{p}_1 = \frac{-c_{22}c_1 + c_{12}c_2}{c_{11}c_{22} - c_{12}c_{21}},\ \hat{p}_2 = \frac{c_{11}c_2 - c_{21}c_1}{c_{11}c_{22} - c_{12}c_{21}},$$

and we have $\hat{p}_1 > 0$ and $\hat{p}_2 > 0$, if and only if

$$c_{11}c_2 - c_{21}c_1 > 0$$

The Jacobi matrix $J_F(\hat{p})$ reads

$$J_F(\hat{p}) = \begin{pmatrix} c_{11}\hat{p}_1 & c_{12}\hat{p}_1 \\ c_{21}\hat{p}_2 & c_{22}\hat{p}_2 \end{pmatrix}.$$

The eigenvalues are given by

$$\lambda_{1,2} = \frac{1}{2}(c_{11}\hat{p}_1 + c_{22}\hat{p}_2) \pm \sqrt{\frac{1}{4}(c_{11}\hat{p}_1 + c_{22}\hat{p}_2)^2 - \underbrace{(c_{11}c_{22} - c_{12}c_{21})}_{>0}\hat{p}_1\hat{p}_2}$$

This implies that $Re(\lambda_1) < 0$ and $Re(\lambda_2) < 0$.

Hence by the above conclusion the eigenvalues of $J_{gh}(\hat{p})$ are smaller than 1 in absolute value, if

$$h < -\frac{2Re(\lambda_i)}{|\lambda_i|^2},\ \text{for } i = 1, 2.$$

We distinguish three cases:

1.
$$\frac{1}{4}(c_{11}\hat{p}_1 + c_{22}\hat{p}_2)^2 = (c_{11}c_{22} - c_{12}c_{21})\hat{p}_1\hat{p}_2.$$

Then it follows that

$$\lambda_1 = \lambda_2 = \frac{1}{2}(c_{11}\hat{p}_1 + c_{22}\hat{p}_2) < 0$$

and we obtain

$$h < \frac{-4}{c_{11}\hat{p}_1 + c_{22}\hat{p}_2}$$

as sufficient condition for the eigenvalues of $J_{gh}(\hat{p})$ being smaller than 1 in absolute value.

2.
$$\frac{1}{4}(c_{11}\hat{p}_1 + c_{22}\hat{p}_2)^2 < (c_{11}c_{22} - c_{12}c_{21})\hat{p}_1\hat{p}_2.$$

Then it follows that
$$Re(\lambda_1) = Re(\lambda_2) = \frac{1}{2}(c_{11}\hat{p}_1 + c_{22}\hat{p}_2) < 0$$
and
$$|\lambda_1|^2 = |\lambda_2|^2 = (c_{11}c_{22} - c_{12}c_{21})\hat{p}_1\hat{p}_2.$$
This implies that
$$h < \frac{-(c_{11}\hat{p}_1 + c_{22}\hat{p}_2)}{(c_{11}c_{22} - c_{12}c_{21})\hat{p}_1\hat{p}_2}$$
is a sufficient condition for the eigenvalues of $J_{gh}(\hat{p})$ being smaller than 1 in absolute value.

3.
$$\frac{1}{4}(c_{11}\hat{p}_1 + c_{22}\hat{p}_2)^2 > (c_{11}c_{22} - c_{12}c_{21})\hat{p}_1\hat{p}_2.$$

Then it follows that $\lambda_1, \lambda_2 \in \mathbb{R}$, $\lambda_2 < \lambda_1 < 0$. Further it is
$$\frac{-2Re(\lambda_i)}{|\lambda_i|^2} = \frac{-2\lambda_i}{\lambda_i^2} = \frac{2}{(-\lambda_i)} \text{ for } i = 1, 2$$
and we obtain
$$h < -\frac{2}{\lambda_2}$$
as sufficient condition for the eigenvalues of $J_{gh}(\hat{p})$ being smaller than 1 in absolute value.

In the case of stable equilibrium states one cannot guarantee that by discretization one obtains stable fixed points, if the time step size is small enough. Let us demonstrate this by the classical Volterra-Lotka-model which is given by
$$\begin{aligned}\dot{p}_1(t) &= (c_1 + c_{12}p_2(t))p_1(t),\\ \dot{p}_2(t) &= (c_2 + c_{21}p_1(t))p_2(t), \ t \in \mathbb{R},\end{aligned} \tag{1.68}$$
where
$$c_1 > 0, \ c_{12} < 0, \ c_2 < 0, \ c_{21} > 0.$$
Here
$$p_1(t) = \hat{p}_1 = -\frac{c_2}{c_{21}}, \ p_2(t) = \hat{p}_2 = -\frac{c_1}{c_{12}}, \text{ for all } t \in \mathbb{R}$$

is the only equilibrium state of (1.68) with $\hat{p}_1 > 0, \hat{p}_2 > 0$. In Section 1.2 we have shown that it is stable.

The discretization of (1.68) is given by

$$\begin{aligned} p_1(t+h) &= (1 + h(c_1 + c_{12}p_2(t)))p_1(t), \\ p_2(t+h) &= (1 + h(c_2 + c_{21}p_1(t)))p_2(t), \quad t \in \mathbb{R}. \end{aligned} \tag{1.69}$$

The Jacobi matrix of

$$g^h(p) = \begin{pmatrix} 1 + h(c_1 + c_{12}p_2)p_1 \\ 1 + h(c_2 + c_{21}p_1)p_1 \end{pmatrix}, \quad p_1 > 0, p_2 > 0,$$

in $\hat{p} = (\hat{p}_1, \hat{p}_2)$ is given by

$$J_{gh}(\hat{p}) = \begin{pmatrix} 1 & -hc_{12}\frac{c_2}{c_{21}} \\ -hc_{21}\frac{c_1}{c_{12}} & 1 \end{pmatrix}$$

and has the eigenvalues

$$\lambda_{1,2} = 1 \pm h\sqrt{|c_1| \cdot |c_2|}i, \quad i = \sqrt{-1},$$

which implies $|\lambda_{1,2}| > 1$ for all $h > 0$. By Satz 1.20 in [4] therefore $(\hat{p}_1, \hat{p}_2)$ is a repelling fixed point of g^h and cannot be stable for sufficiently small step size $h > 0$.

Now let us replace the system (1.69) by

$$\begin{aligned} p_1(t+h) &= (1 + h(c_1 + c_{12}p_2(t)))p_1(t), \\ p_2(t+h) &= (1 + h(c_2 + c_{21}p_1(t+h)))p_2(t), \quad t \in \mathbb{R}, \end{aligned} \tag{1.70}$$

and define a vector function $\tilde{g}^h : \overset{\circ}{\mathbb{R}}{}^2_+ \to \mathbb{R}^2$ by

$$\tilde{g}^h(p) = \begin{pmatrix} (1 + h(c_1 + c_{12}p_2))p_1 \\ (1 + h(c_2 + c_{21}(p_1 + h(c_1 + c_{12}p_2)p_1))p_2 \end{pmatrix}, \quad p_1 > 0, p_2 > 0.$$

Then $(\hat{p}_1, \hat{p}_2)$ is also a fixed point of $\tilde{g}^h$ and the question arises whether it is stable for sufficiently small $h > 0$.

In order to find an answer to this question we consider the Jacobi matrix of $\tilde{g}^h$ in $(\hat{p}_1, \hat{p}_2)$ which is given by

$$J_{\tilde{g}^h}(\hat{p}) = \begin{pmatrix} 1 & -hc_{12}\frac{c_2}{c_{21}} \\ -hc_{21}\frac{c_1}{c_{12}} & 1 + h^2c_1c_2 \end{pmatrix}. \tag{1.71}$$

It has the eigenvalues

$$\lambda_{1,2} = 1 + \frac{h^2}{2}c_1c_2 \pm \sqrt{\left(1 + \frac{h^2}{2}c_1c_2\right)^2 - 1}.$$

Now we distinguish three cases:

1.
$$\left(1 + \frac{h^2}{2}c_1c_2\right)^2 = 1 \iff |c_1c_2| = \frac{4}{h^2}.$$

   Then it follows that

$$\lambda_1 = \lambda_2 = 1 + \frac{h^2}{2}c_1c_2 = 1.$$

2.
$$\left(1 + \frac{h^2}{2}c_1c_2\right)^2 < 1 \iff |c_1c_2| < \frac{4}{h^2}.$$

   Then it follows that

$$\lambda_{1,2} = 1 + \frac{h^2}{2}c_1c_2 \pm i\sqrt{1 - \left(1 + \frac{h^2}{2}c_1c_2\right)^2}, \quad i = \sqrt{-1},$$

   which implies that

$$|\lambda_1| = |\lambda_2| = 1 \text{ and } \lambda_2 = \overline{\lambda_1} \neq \lambda_1.$$

3.
$$\left((1 + \frac{h^2}{2}c_1c_2\right)^2 > 1 \iff |c_1c_2| > \frac{4}{h^2}.$$

   Then it follows that

$$\begin{aligned}\lambda_{1,2} &= 1 + \frac{h^2}{2}c_1c_2 \pm \sqrt{h^2c_1c_2 + \frac{h^4}{4}c_1^2c_2^2} \\ &= 1 + \frac{h^2}{2}c_1c_2 \pm \frac{h^2}{2}|c_1c_2|\sqrt{1 - \frac{4}{h^2|c_1c_2|}}\end{aligned}$$

   and hence

$$\lambda_2 < \lambda_1 < 1 + \frac{h^2}{2}c_1c_2 + \frac{h^2}{2}|c_1c_2| = 1.$$

Now let us assume that $\lambda_2 \geq -1$. Then it follows

$$|c_1c_2|\left(1+\sqrt{1-\frac{4}{h^2|c_1c_2|}}\right) \leq \frac{4}{h^2}$$

which contradicts $|c_1c_2| > \frac{4}{h^2}$. Therefore we conclude that $\lambda_2 < -1$. If we linearize the system (1.70) around $\hat{p}$ we obtain the system

$$\begin{pmatrix} u_1(t+h) \\ u_2(t+h) \end{pmatrix} = J_{\tilde{g}^h}(\hat{p}) \begin{pmatrix} u_1(t) \\ u_2(t) \end{pmatrix} \tag{1.72}$$

with $J_{\tilde{g}^h}(\hat{p})$ given by (1.71).

Now we assume $|c_1c_2| < \frac{4}{h^2}$ (which can be achieved, if $h > 0$ is chosen sufficiently small). We have seen above that this implies

$$\lambda_2 = \overline{\lambda_1} \neq \lambda_1 \text{ and } |\lambda_1| = |\lambda_2| = 1.$$

By Theorem 1.7 in [8], $(0,0)$ is a stable fixed point solution of the system (1.72). In this sense the fixed point $\hat{p}$ of $\tilde{g}^h$ is stable for sufficiently small $h > 0$. If $c_{22} < 0$ and $c_{11} < 0$ the discretization (1.69) of (1.68) has to be replaced by

$$\begin{aligned} p_1(t+h) &= (1 + h(c_1 + c_{11}p_1(t) + c_{12}p_2(t)))p_1(t), \\ p_2(t+h) &= (1 + h(c_2 + c_{21}p_1(t) + c_{22}p_2(t)))p_2(t), \quad t \in \mathbb{R}. \end{aligned}$$

We have shown above that the eigenvalues of $J_{\tilde{g}^h}(\hat{p})$ are smaller than 1 in absolute value (which implies that the fixed point $\hat{p}$ of the righthand side is attractive), if

$$h < \frac{|a|}{b} \text{ in case } \frac{a^2}{4} < b$$

where

$$a = c_{11}\hat{p}_1 + c_{22}\hat{p}_2 \text{ and } b = (c_{11}c_{22} - c_{12}c_{21})\hat{p}_1\hat{p}_2$$

and

$$h < \frac{4}{|a| + \sqrt{a^2 - 4b}} \text{ in case } \frac{a^2}{4} \geq b.$$

In analogy to (1.70) let us now replace the above discretization by

$$\begin{aligned} p_1(t+h) &= (1 + h(c_1 + c_{11}p_1(t) + c_{12}p_2(t)))p_1(t), \\ p_2(t+h) &= (1 + h(c_2 + c_{21}p_1(t+h) + c_{22}p_2(t)))p_2(t), \quad t \in \mathbb{R}. \end{aligned}$$

Then the righthand side $\tilde{g}^h$ has the same fixed point $\hat{p} \in \mathbb{R}^2_+$ as g^h and the Jacobi matrix $J_{\tilde{g}^h}(\hat{p})$ of $\tilde{g}^h$ in $\hat{p}$ is given by

$$J_{\tilde{g}^h}(\hat{p}) = \begin{pmatrix} 1 + hc_{11}\hat{p}_1 & hc_{12}\hat{p}_1 \\ hc_{21}\hat{p}_2 & h^2 c_{21}c_{12}\hat{p}_1\hat{p}_2 + 1 + hc_{22}\hat{p}_2 \end{pmatrix}.$$

The eigenvalues of $J_{\tilde{g}^h}(\hat{p})$ are solutions of the quadratic equation

$$\mu^2 - (2 + ha + h^2 c)\mu + 1 + ha + h^2(b + c) = 0$$

where

$$c = c_{12}c_{21}\hat{p}_1\hat{p}_2 < 0.$$

They are given by

$$\begin{aligned} \mu_{1,2} &= 1 + \frac{h}{2}(a + hc) \pm \sqrt{\left(1 + \frac{h}{2}(a + hc)\right)^2 - 1 - ha - h^2(b + c)} \\ &= 1 + \frac{h}{2}(a + hc) \pm \frac{h}{2}\sqrt{(a + hc)^2 - 4b}. \end{aligned}$$

We distinguish two cases:

1. $(a + hc)^2 < 4b$, then it follows that

$$|\mu_1|^2 = |\mu_2|^2 = 1 + ha + h^2(b + c)$$

and

$$|\mu_1|^2 = |\mu_2|^2 < 1$$

is equivalent to

$$h < \frac{|a|}{b + c} \quad (b + c = c_{11}c_{22}\hat{p}_1\hat{p}_2 > 0).$$

Since $(a + hc)^2 = (|a| + h|c|)^2 < 4b$ implies $a^2 < 4b$ in which case the eigenvalues of $J_{gh}(\hat{p})$ are smaller than 1 in absolute value, if

$$h < \frac{|a|}{b}.$$

Because of $0 < b + c < b$ therefore the attractivity of the fixed point $\hat{p}$ of $\tilde{g}^h$ is possible for a larger step size then for g^h.

2. $(a + hc)^2 \geq 4b$, then it follows that

$$\mu_2 < \mu_1 < 1 + h(a + hc) < 1$$

and

$$\mu_2 = 1 + \frac{h}{2}(a + hc) - \frac{h}{2}\sqrt{(a + hc)^2 - 4b} > -1$$

is equivalent to

$$h \leq \frac{4}{|a| + h|c| + \sqrt{(a + hc)^2 - 4b}}.$$

Now $a^2 \geq 4b$ implies $(a + hc)^2 \geq 4b$ for every $h > 0$.

Because of

$$\frac{4}{|a| + \sqrt{a^2 - 4b}} > \frac{4}{|a| + h|c| + \sqrt{(a + hc)^2 - 4b}}$$

therefore the attractivity of the fixed point $\hat{p}$ of g^h is possible for a larger size than for $\tilde{g}^h$.

Finally let us consider the case $c_{11} = 0$ and $c_{22} < 0$. Then it follows that $b + c = 0$ and therefore

$$\begin{aligned}\mu_{1,2} &= 1 + \frac{h}{2}(a + hc) \pm \sqrt{(1 + \frac{h}{2}(a + hc))^2 - 1 - ha} \\ &= 1 + \frac{h}{2}(a + hc) \pm \frac{h}{2}\sqrt{(a + hc)^2 + c}.\end{aligned}$$

Further we have

$$a = c_{22}\hat{p}_2 < 0, \; b = -c = -c_{12}c_{21}\hat{p}_1\hat{p}_2 > 0.$$

We again distinguish two cases:

1. $(a + hc)^2 < 4b$, then it follows that $1 + ha > 0$ and

$$|\mu_1|^2 = |\mu_2|^2 = 1 + ha < 1$$

which is equivalent to

$$h < \frac{1}{|a|}.$$

Hence the fixed point $\hat{p}$ of $\tilde{g}^h$ is attractive, if $h < \frac{1}{|a|}$.

We have shown above that the same fixed point $\hat{p}$ of g^h is attractive, if $h < \frac{|a|}{b}$. So it follows that the attractivity of $\hat{p}$ for $\tilde{g}^h$ is possible for larger h then for g^h, if

$$\frac{|a|}{b} < \frac{1}{|a|} \iff a^2 < b.$$

2. $(a + hc)^2 \geq 4b$, here we have the same conclusion as in the case $a_{11} < 0$ and $a_{22} < 0$.

1.4.2 The One-Population Model

For $n = 1$ we start with the equation (1.14) as model for the growth of one population, i.e. with the equation

$$\dot{p}(t) = f(p(t))p(t), \ t \in \mathbb{R}, \tag{1.73}$$

where we assume $f : \mathbb{R} \to \mathbb{R}$ to be continuously differentiable. The discretization of this equation is then given by

$$p(t + h) = (1 + hf(p(t)))p(t), \ t \in \mathbb{R}, \tag{1.74}$$

where $h > 0$ is a time step size. If $\hat{p} > 0$ is a solution of the equation $f(\hat{p}) = 0$ (i.e. $p(t) = \hat{p}, t \in \mathbb{R}$, is an equilibrium state of the equation (1.73)), then $\hat{p}$ is a fixed point of the function

$$g^h(p) = (1 + hf(p))p, \ p \in \mathbb{R},$$

i.e., a solution of the equation $g^h(\hat{p}) = \hat{p}$, and vice versa.

The derivative of g^h in $\hat{p}$ is given by

$$\frac{dg^h}{dp}(\hat{p}) = hf'(\hat{p})\hat{p} + 1$$

from which we infer that

$$\left|\frac{dg^h}{dp}(\hat{p})\right| < 1,$$

if and only if

$$f'(\hat{p}) < 0 \text{ and } h < \frac{2}{-f'(\hat{p})\hat{p}}. \tag{1.75}$$

This implies that $\hat{p} > 0$ is an attractive fixed point of g^h, i.e., there exists a neighbourhood $\mathcal{U}(\hat{p}) \subseteq \overset{\circ}{\mathbb{R}}{}^n$ of $\hat{p}$ such that

$$\lim_{h\to\infty} (g^h)^k(p) = \hat{p} \text{ for all } p \in \mathcal{U}(\hat{p}), \tag{1.76}$$

if $f'(\hat{p}) < 0$ and $h > 0$ is sufficiently small. From (1.75) we even get an estimate for h.

Let us demonstrate this by the second growth model of Verhulst in which f is given by the linear function

$$f(p) = a - bp, \;\; p \in \mathbb{R},$$

with $a > 0$ and $b > 0$ being given constants.

In this case we have

$$f'(p) = -b < 0 \text{ for all } p \in \mathbb{R} \text{ and } \hat{p} = \frac{a}{b}.$$

Therefore $\hat{p}$ is an attractive fixed point of

$$g^h(p) = (1 + h(a - bp))p, \;\; p \in \mathbb{R},$$

if $h < \frac{2}{a}$. In this case we can give an explicit representation of $\mathcal{U}(\hat{p})$ in (1.76). For this purpose we define

$$r = ah \text{ and } q = bh.$$

Then g^h can be written in the form (with $x := p$)

$$g(x) = (1 + r)x - qx^2 = (1 + r)x\left(1 - \frac{q}{r+1}x\right), \;\; x \in \mathbb{R}.$$

If we put $\underline{\overline{X}} = \left[0, \frac{r+1}{q}\right]$ then it follows that

$$g(\underline{\overline{X}}) \subseteq \underline{\overline{X}}, \text{ if } 0 \le r \le 3.$$

Further we obtain

$$g(\hat{x}) = \hat{x} > 0 \iff \hat{x} = \frac{r}{q}.$$

This fixed point of g (which is also a fixed point of g^h) is attractive, if $r \in (0, 2)$ ($\iff 0 < h < \frac{2}{a}$).

Graphically we have the situation illustrated by Figure 1.4 and Figure 1.5.

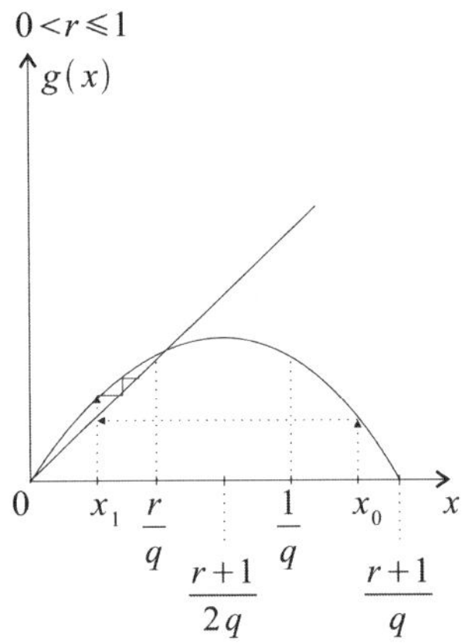

Fig. 1.4. Discrete Second Verhulst Model a)

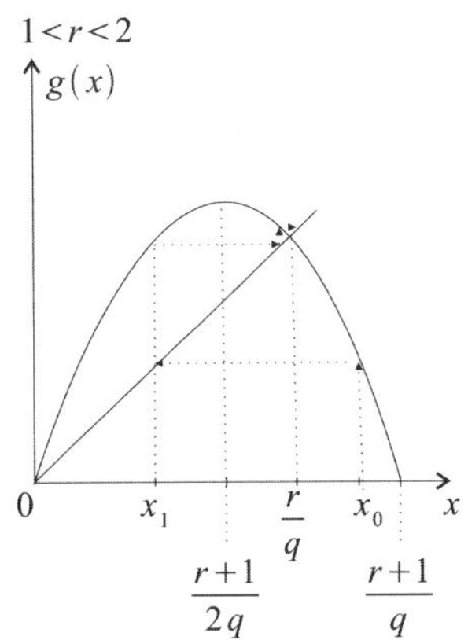

Fig. 1.5. Discrete Second Verhulst Model b)

In case a) every sequence $(x_{k+1} = g(x_k))_{k\in\mathbb{N}_0}$ converges monotonically increasing after x_0 or x_1 to $\hat{x} = \frac{r}{q}$, if x_0 is chosen in $\left(0, \frac{r}{q}\right) \cup \left(\frac{1}{q}, \frac{r+1}{q}\right)$, and monotonically decreasing to $x = \frac{r}{q}$, if x_0 is chosen in $\left(\frac{r}{q}, \frac{1}{q}\right)$.

In case b) every sequence $(x_{k+1} = g(x_k))_{k\in\mathbb{N}_0}$ converges for every choice of $x_0 \in \left(0, \frac{r+1}{q}\right)$ either from the beginning or after finitely many steps alternatingly to $\hat{x} = \frac{r}{q}$.

Result: If $r \in (0, 2)$, every sequence $(x_{k+1} = g(x_k))_{k\in\mathbb{N}_0}$ converges to $\hat{x} = \frac{r}{q}$, if x_0 is chosen in $\left(0, \frac{r+1}{q}\right)$.

In other words: If $0 < h < \frac{2}{a}$, then the neighbourhood $\mathcal{U}(\hat{p})$ in (1.76) can be chosen as $\left(0, \frac{r+1}{q}\right)$.

1.5 Determination of Model Parameters from Data

We start with the classical Volterra-Lotka-model which is described by the differential equations (1.68).

Let us assume that we have given data $(p_1(i \cdot h), p_2(i \cdot h))$ for some $h > 0$ and $i = 0, \ldots, N$. The question now is whether and how these data can be fitted into the model (1.68). In order to give a reasonable answer to this question we replace the system (1.68) by its modified discretization (1.70) and ask whether the given data can be fitted into this model. For that purpose we rewrite (1.70) in the form

$$\frac{p_1(t+h)-p_1(t)}{h} = c_1 p_1(t) + c_{12} p_1(t) p_2(t),$$

$$\frac{p_2(t+h)-p_2(t)}{h} = c_2 p_2(t) + c_{21} p_1(t+h) p_2(t)$$

The coefficients c_1, c_{12}, c_2, c_{21} are then determined so that

$$\sum_{i=0}^{N-1} \left(\frac{p_1((i+1)h) - p_1(i \cdot h)}{h} - c_1 p_1(i \cdot h) - c_{12} p_1(i \cdot h) p_2(i \cdot h) \right)^2 \quad (1.77)$$

and

$$\sum_{i=0}^{N-1} \left(\frac{p_2((i+1)h) - p_2(i \cdot h)}{h} - c_2 p_2(i \cdot h) - c_{21} p_1((i+1) \cdot h) p_2((i \cdot h) \right)^2 \quad (1.78)$$

become as small as possible.

If the result of this minimisation is given by

$$c_1 > 0,\ c_{12} < 0,\ c_2 < 0,\ c_{21} > 0 \text{ or } c_1 < 0,\ c_{12} > 0,\ c_2 > 0,\ c_{21} < 0,$$

then we accept the data as being compatible with the model (1.68).

In the first case the population P_1 is the prey and P_2 the predator and in the second case vice versa.

Let us demonstrate this by an example. The data of this example are taken from [9]. We present them in the following table:

$t_i = i$	0	1	2	3	4	5	6	7	8	$h = 1$
$p_1(t_i)$	0.300	0.148	0.144	0.248	0.171	0.284	0.243	0.191	0.166	
$p_2(t_i)$	0.688	0.851	0.852	0.707	0.788	0.624	0.679	0.761	0.781	

By minimizing (1.77) and (1.78) we obtain

$$c_1 = -2.34,\ c_{12} = 3.09,\ c_2 = 0.42,\ c_{21} = -2.08$$

hence

$$\hat{p}_1 = \frac{-c_2}{c_{21}} = 0.202,$$
$$\hat{p}_2 = -\frac{c_1}{c_{12}} = 0.757.$$

Further we get

$$|c_1 \cdot c_2| = 0.9828 < \frac{4}{h^2} = 4.$$

Next we consider a modification of the Volterra-Lotka-model which is described by the system

$$\begin{aligned} \dot{p}_1(t) &= (c_1 + c_{12}p_2(t))p_1(t), \\ \dot{p}_2(t) &= (c_2 + c_{21}p_1(t) + c_{22}p_2(t))p_2(t),\ t \in \mathbb{R}, \end{aligned} \tag{1.79}$$

where

$$c_1 < 0,\ c_2 > 0,\ c_{12} > 0,\ c_{21} < 0,\ c_{22} < 0 \tag{1.80}$$

(i.e. P_1 is the predator and P_2 is the prey population). If we define

$$\hat{p}_1 = -\frac{1}{c_{21}}\left(c_2 - c_{22}\frac{c_1}{c_{12}}\right),$$
$$\hat{p}_2 = -\frac{c_1}{c_{12}}$$

then

$$p_1(t) = \hat{p}_1,\ p_2(t) = \hat{p}_2 \text{ for all } t \in \mathbb{R}$$

is an equilibrium state of (1.79) with $\hat{p}_1 > 0$ and $\hat{p}_2 > 0$, if the condition

$$c_{12}c_2 - c_{22}c_1 > 0$$

is satisfied. In [4] we have shown that this equilibrium state is asymptotically stable.

For a given time step size $h > 0$ the discretization of the system (1.79) can be written in the form

$$\frac{p_1(t+h) - p_1(t)}{h} = c_{12}(p_2(t) - \hat{p}_2)p_1(t)$$
$$\frac{p_2(t+h) - p_2(t)}{h} = c_2 p_2(t) + c_{21}p_1(t) + c_{22}p_2(t)^2.$$

In order to determine c_{12} we choose some $\hat{p}_2$, say

$$\hat{p}_2 = \frac{1}{N+1} \sum_{i=0}^{N} p_2(i \cdot h),$$

if again $(p_1(i \cdot h), p_2(i \cdot h)),\ i = 0, \dots, N$, are given data, and minimize

$$\sum_{i=0}^{N-1} \left(\frac{p_1((i+1)h) - p_1(i \cdot h)}{h} - (p_2(i \cdot h) - \hat{p}_2)\, p_1(i \cdot h) c_{12} \right)^2 . \tag{1.81}$$

For the data given in the above table we chose $\hat{p}_2 = 0.75$ and obtain by minimizing (1.81)

$$c_{12} = 3.23 \text{ and } c_1 = -2.42.$$

In order to determine $c_2,\ c_{21},\ c_{22}$ we minimize

$$\sum_{i=0}^{N-1} \Big(\frac{p_2((i+1)h) - p_2(i \cdot h)}{h}$$
$$-p_2(i \cdot h)c_2 - p_1(i \cdot h)p_2(i \cdot h)c_{21} + p_2(i \cdot h)^2 c_{22} \Big)^2 . \tag{1.82}$$

For the data given in the above table, however, we obtain the values

$$c_2 = -1.62,\ c_{21} = 3.6,\ c_{22} = 1.16$$

which violate the condition (1.80).

If we, however, replace (1.82) by

$$\sum_{i=0}^{N-1} \Big(\frac{p_2((i+1)h) - p_2(i \cdot h)}{h}$$
$$-p_2(i \cdot h)c_2 - p_1((i+1)h)p_2(i \cdot h)c_{21} + p_2(i \cdot h)^2 c_{22} \Big)^2 , \tag{1.83}$$

then we obtain from the above data

$$c_2 = 1.3,\ c_{21} = -1.92,\ c_{22} = -1.2$$

which leads to $\hat{p}_1 = 0.21$. Further we obtain with

$$\begin{aligned} a &= c_{22}\hat{p}_2 = -0.9, \\ b &= -c_{12}c_{21}\hat{p}_1\hat{p}_2 = 0.977, \\ a + c &= -1.877 \end{aligned}$$

that

$$(a+c)^2 = 3.522 < 4b = 3.908.$$

Because of $\frac{1}{|a|} = 1.\overline{1}\ldots > 1$ we conclude that $\hat{p} = (0.21, 0.75)$ is an attractive fixed point (see Section 1.4.1).

References

[1] M. Braun: Differentialgleichungen und ihre Anwendungen. Springer-Verlag: Berlin-Heidelberg-New York 1979.

[2] M. Eisen: Mathematical Models in Biology and Cancer Chemotherapy. Lecture Notes in Biomathematics. Springer-Verlag: Berlin-Heidelberg-New York 1979.

[3] A. Hurwitz: Über die Bedingungen unter welchen eine Gleichung nur Wurzeln mit negativen reellen Teilen besitzt. Math. Ann. Bd. 46 (1895), 273-284.

[4] W. Krabs: Dynamische Systeme: Steuerbarkeit und chaotisches Verhalten. Verlag B.G. Teubner: Stuttgart-Leipzig 1998.

[5] W. Krabs: A General Predator-Prey Model. Mathematical and Computer Modelling of Dynamical Systems. 9 (2003), 387-401.

[6] W. Krabs: Stability in Predator-Prey Models and Discretization of a Modified Volterra-Lotka-Model. Mathematical and Computer Modelling of Dynamical Systems. 12 (2006), 577-588.

[7] W. Krabs: Spieltheorie: Dynamische Behandlung von Spielen. Verlag B.G. Teubner: Stuttgart-Leipzig-Wiesbaden, 2005.

[8] W. Krabs: and S. W. Pickl: Analysis, Controllability and Optimisation of Time-Discrete Systems and Dynamical Games. Lecture Notes in Economics and Mathematical Systems. Springer-Verlag Berlin-Heidelberg, 2003.

[9] W.Krabs and R. Simon: Räuber-Beute-Verhalten in kleinräumigen Habitaten. Manuskript.

2

A Game-Theoretic Evolution Model

2.1 Evolution-Matrix-Games for one Population

2.1.1 The Game and Evolutionarily Stable Equilibria

During the last thirty years, based on a paper of the biologist J. Maynard Smith with the title "Game Theory and the Evolution of Fighting", a theory of evolution games has been developed. This starts with a population of individuals who have a finite number $I_1, I_2, \ldots, I_n$ of strategies in order to survive in the struggle of life. Let $u_i = [0, 1]$, for every $i = 1, \ldots, n$ be the probability for the strategy I_i to be chosen in the population. Then the corresponding state of the population is defined by the vector $u = (u_1, \ldots, u_n)$ where $\sum\limits_{i=1}^{n} u_i = 1$. The set of all population states is given by the simplex

$$\Delta = \left\{ u = (u_1, \ldots, u_n) | 0 \leq u_i \leq 1, i = 1, \ldots, n, \sum_{i=1}^{n} u_i = 1 \right\}.$$

Every vector $e_i = (0, \ldots, 0, 1_i, 0, \ldots, 0)$, $i = 1, \ldots, n$, denotes a so called pure population state where all the individuals choose the strategy I_i. All the other states are called mixed states. If an individual that chooses strategy I_i meets an individual that chooses strategy I_j, we assume that the I_i-individual is given a payoff $a_{ij} \in \mathbb{R}$ by the I_j-individual. All the payoffs then form a matrix

$$A = (a_{ij})_{i,j=1,\ldots,n}$$

the so called payoff matrix which defines a matrix game. This, however, is in general not a zero-sum game with $A = -A^T$ (see Section 2.1.7).

The expected payoff of an I_i-individual in the population state $u \in \Delta$ is defined by

$$\sum_{j=1}^{n} a_{ij} u_j = e_i A u^T.$$

If two populations states $u, v \in \Delta$ are given, then the average payoff of u to v is defined by

$$\sum_{i,j=1}^{n} a_{ij} u_i v_j = v A u^T.$$

Definition. *A population state $u^* \in \Delta$ is called a Nash equilibrium if*

$$uAu^{*T} \le u^* A u^{*T}, \text{ for all } u \in \Delta.$$

In words this means that a deviation from u^* does not lead to a higher payoff. For rational behavior this would suffice to maintain the population state u^*. However, animals do not behave rationally so that the stability of a Nash equilibrium is not guaranteed. This leads to the concept of evolutionarily stable Nash equilibrium given by the

Definition. *A Nash equilibrium $u^* \in \Delta$ is called evolutionarily stable, if $uAu^{*T} = u^* A u^{*T}$ for some $u \in \Delta$ with $u \neq u^*$ implies that $uAu^T < u^* A u^T$. In words this means that, if a change from u^* to u leads to the same payoff, u cannot be a Nash equilibrium.*

Let us demonstrate these definitions by an example. We consider a population with two strategies I_1 and I_2. Individuals that choose I_1 are called pigeons and those who choose I_2 are called hawks. If a pigeon meets a pigeon they menace each other without seriously fighting until one of them gives in. If a pigeon meets a hawk, it runs away and is not hurt. If two hawks meet each other they fight until one of them is seriously hurt and has to give up or is dead. Let us assume that the winner is given $V > 0$ points and the loser in a fight of hawks is given $-D$ points where $D > 0$. This leads to the payoff matrix

$$A = \begin{pmatrix} \frac{V}{2} & 0 \\ V & \frac{V-D}{2} \end{pmatrix}.$$

Let us at first assume that $V \ge D$.

Then we assert that the pure population state $e_2 = (0, 1)$ is an evolutionarily stable Nash equilibrium.

In order to show that we choose an arbitrary $u \in \Delta$ and find that

$$e_2 A e_2^T - u A e_2^T = \frac{V-D}{2}(1-u_2) \geq 0$$

which shows that e_2 is a Nash equilibrium.

If $uAe_2^T = e_2 A e_2^T$, then it follows that $\frac{V-D}{2}(1-u_2) = 0$.

If $V > D$, then it follows that $u_2 = 1$ and $u_1 = 0$, hence $u = e_2$.

If $V = D$, then it follows that $uAe_2^T = e_2 A e_2^T$ for all $u \in \Delta$, and for all $u \in \Delta$ with $u \neq e_2$ it follows that

$$uAu^T = \frac{V}{2} u_1 (1-u_2) < V u_1 = e_2 A u^T$$

which implies that e_2 is evolutionarily stable.

Result. *If $V \geq D$ and if all individuals behave like hawks, then this state is an evolutionarily stable Nash equilibrium.*

On the other hand, if $V < D$, then e_2 is not even a Nash equilibrium. On the contrary we have

$$e_2 A e_2^T - u A e_2^T = \frac{V-D}{2}(1-u_2) < 0 \text{ for all } u \in \Delta \text{ with } u_2 < 1.$$

But also the pure population state $e_1 = (1, 0)$ is not a Nash equilibrium for we have

$$e_1 A e_1^T - u A e_1^T = -\frac{V}{2}(1-u_1) < 0 \text{ for all } u \in \Delta \text{ with } u_1 < 1.$$

The case $V \geq D$ is a special case of the following situation:

Let for some $k \in \{1, \ldots, n\}$

$$a_{kk} \geq a_{jk} \text{ for all j=1, \ldots, n}$$

and

$$a_{kk} = a_{jk} \implies a_{ki} > a_{ji} \text{ for all } i \neq k.$$

Then it follows for every $u \in \Delta$ that

$$uAe_k^T = \sum_{j=1}^{n} u_j a_{jk} \leq \left(\sum_{j=1}^{n} u_j \right) a_{kk} = a_{kk} = e_k A e_k^T,$$

i.e., e_k is a Nash equilibrium.

Now let $u \in \Delta$ with $u \neq e_k$ and $uAe_k^T = e_kAe_k^T$ be given.

Then it follows that

$$a_{jk} = a_{kk} \text{ for all } j \text{ with } u_j > 0$$

and hence

$$a_{ki} > a_{ji} \text{ for all } j \text{ with } u_j > 0 \text{ and all } i \neq k.$$

This implies

$$\begin{aligned} e_kAu^T - uAu^T &= \sum_{i=1}^{n} a_{ki}u_i - \sum_{j=1}^{n}\sum_{i=1}^{n} a_{ji}u_ju_i \\ &= \sum_{j=1}^{n}\sum_{i=1}^{n}(a_{ki} - a_{ji})u_ju_i \\ &= \sum_{u_j>0}\sum_{i\neq k}(a_{ki} - a_{ji})u_ju_i > 0 \end{aligned}$$

which shows that e_k is evolutionarily stable.

Assertion. *If $V < D$, the population state $\left(1 - \frac{V}{D}, \frac{V}{D}\right)$ is an evolutionarily stable Nash equilibrium.*

Proof. If we put $u^* = \left(1 - \frac{V}{D}, \frac{V}{D}\right)$, then $u^* \in \Delta$ and

$$\begin{aligned} u^*Au^{*T} &= \frac{V}{2}\left(1 - \frac{V}{D}\right)^2 - \frac{V^2}{2D}\left(1 - \frac{V}{D}\right) \\ &= \frac{V}{2}\left(1 - \frac{V}{D}\right)\left(1 - \frac{V}{D} + \frac{V}{D}\right) \\ &= \frac{V}{2}\left(1 - \frac{V}{D}\right) \end{aligned}$$

Further it follows for every $u \in \Delta$ that

$$uAu^{*T} = (u_1, u_2)\begin{pmatrix} \frac{V}{2}\left(1 - \frac{V}{D}\right) \\ \frac{V}{2}\left(1 - \frac{V}{D}\right) \end{pmatrix} = \frac{V}{2}\left(1 - \frac{V}{D}\right) = u^*Au^{*T},$$

i.e., u^* is a Nash equilibrium. □

Now we have, for every $u \in \Delta$,

$$\begin{aligned} uAu^T &= \frac{V}{2}u_1^2 + Vu_1u_2 + \frac{V-D}{2}u_2^2 \\ &= \frac{V}{2}\underbrace{(u_1^2 + 2u_1u_2 + u_2^2)}_{=(u_1+u_2)^2=1} - \frac{D}{2}u_2^2 = \frac{V}{2}\left(1 - \frac{D}{V}u_2^2\right) \end{aligned}$$

and

$$\begin{aligned} u^*Au^T &= \left(1 - \frac{V}{D}, \frac{V}{D}\right)\begin{pmatrix} \frac{V}{2}u_2 \\ Vu_1 + \frac{V-D}{2}u_2 \end{pmatrix} \\ &= \frac{V}{2}\left(1 - \frac{V}{D}\right)u_1 + \frac{V}{D}\left(Vu_1 + \frac{V-D}{2}u_2\right) \\ &= \frac{V}{2}u_1 - \frac{V^2}{2D}u_1 + \frac{V^2}{D}u_1 + \frac{V^2}{2D}u_2 - \frac{V}{2}u_2 \\ &= \frac{V}{2}u_1 + \frac{V^2}{2D} - \frac{V}{2}u_2 \\ &= \frac{V}{2} + \frac{V^2}{2D} - Vu_2. \end{aligned}$$

Therefore it follows that, for every $u \in \Delta$ with $u \neq u^*$,

$$\begin{aligned} uAu^T - u^*Au^T &= \frac{V}{2} - \frac{D}{2}u_2^2 - \frac{V}{2} - \frac{V^2}{2D} + Vu_2 \\ &= -\frac{D}{2}\left(u_2^2 - \frac{2V}{D}u_2 + \frac{V^2}{D^2}\right) \\ &= -\frac{D}{2}\left(u_2 - \frac{V}{D}\right)^2 < 0, \end{aligned}$$

i.e., u^* is evolutionarily stable.

2.1.2 Characterization of Evolutionarily Stable Equilibria

We begin with a necessary condition for a Nash equilibrium. For this purpose we define for every $u \in \Delta$ a support by

$$S(u) = \{i \in \{1, \ldots, n\} | u_i > 0\}.$$

Then we can prove

Lemma 1. *If $u^* \in \Delta$ is a Nash equilibrium, then*

$$e_iAu^{*T} = u^*Au^{*T} \text{ for all } i \in S(u^*). \tag{2.1}$$

Proof. At first we have

$$u^*Au^{*T} = \sum_{i=1}^{n} u_i^* e_i A u^{*T} = \sum_{i \in S(u^*)} u_i^* e_i A u^{*T} \leq \max_{i \in S(u^*)} e_i A u^{*T}.$$

Since u^* is a Nash equilibrium, it follows that

$$e_i A u^{*T} \leq u^* A u^{*T} \text{ for all } i \in S(u^*),$$

hence

$$u^* A u^{*T} = \max_{i \in S(u^*)} e_i A u^{*T}.$$

□

Let

$$e_{i_0} A u^{*T} = \max_{i \in S(u^*)} e_i A u^{*T}.$$

Then we obtain

$$0 = u^* A u^{*T} - \sum_{i \in S(u^*)} u_i^* e_i A u^{*T} = \sum_{i \in S(u^*)} u_i^* (e_{i_0} A u^{*T} - e_i A u^{*T})$$

which implies

$$u^* A u^{*T} = e_{i_0} A u^{*T} = e_i A u^{*T} \text{ for all } i \in S(u^*).$$

As an immediate consequence we get the

Corollary 1. *If $u^* \in \Delta$ is a Nash equilibrium, then*

$$uAu^{*T} = u^*Au^{*T} \textit{ for all } u \in \Delta \textit{ with } S(u) \subseteq S(u^*). \tag{2.2}$$

Proof. Now we have for every $u \in \Delta$ with $S(u) \subseteq S(u^*)$, because of (2.1),

$$uAu^{*T} = \sum_{i \in S(u)} u_i e_i A u^{*T} = \sum_{i \in S(u)} u_i (u^* A u^{*T}) = u^* A u^{*T}.$$

□

From this we obtain immediately the

Corollary 2. *If $u^* \in \Delta$ is a Nash equilibrium with*

$$u_i^* > 0 \textit{ for all } i \in \{1, \ldots, n\} \iff S(u^*) = \{1, \ldots, n\}, \tag{2.3}$$

then

$$uAu^{*T} = u^*Au^{*T} \textit{ for all } u \in \Delta. \tag{2.4}$$

From Corollary 2 we deduce the

Theorem 2.1 *If $u^* \in \Delta$ is an evolutionarily stable Nash equilibrium with (2.3), then u^* is the only Nash equilibrium.*

Proof. Given an arbitrary $u \in \Delta$ with $u \neq u^*$ it follows from Corollary 2 that $uAu^{*T} = u^*Au^{*T}$. Since u^* is evolutionarily stable, it follows that $uAu^T < u^*Au^T$ so that u cannot be a Nash equilibrium. Therefore u^* is the only Nash equilibrium. □

If the condition (2.3) is not satisfied, then one can show that an evolutionarily stable Nash equilibrium $u^* \in \Delta$ is the only Nash equilibrium in a neighbourhood of u^*.

In order to show that we need some preparations: Let $u, u^* \in \Delta$ be given with $u \neq u^*$ and let $\epsilon \in (0, 1]$. Then we define $w_\epsilon = (1-\epsilon)u^* + \epsilon u$ and conclude that

$$w_\epsilon A w_\epsilon^T = (1-\epsilon)u^* A w_\epsilon^T + \epsilon u A w_\epsilon^T.$$

From this we obtain the equivalence

$$w_\epsilon A w_\epsilon^T < u^* A w_\epsilon^T \iff u A w_\epsilon^T < u^* A w_\epsilon^T. \tag{2.5}$$

Now let $u^* \in \Delta$ be an evolutionarily stable Nash equilibrium and let $u \in \Delta$ be chosen arbitrarily. Then we have

$$uAu^{*T} \leq u^*Au^{*T}.$$

1. Assume that

   $$uAu^{*T} < u^*Au^{*T} \text{ and } u \neq u^*.$$

   Then there is a relatively open set $V_u \subseteq \Delta$ with $u^* \in V_u$ such that

   $$uAv^T < u^*Av^T \text{ for all } v \in V_u \text{ with } v \neq u^*.$$

   Now there exists some $\epsilon_u > 0, \epsilon_u \leq 1$ such that

   $$w_\epsilon = (1-\epsilon)u^* + \epsilon u \in V_u \text{ for all } \epsilon \in [0, \epsilon_u].$$

   This implies

   $$uAw_\epsilon^T < u^*Aw_\epsilon^T.$$

   Using the above equivalence (2.5) we obtain

   $$w_\epsilon A w_\epsilon^T < u^* A w_\epsilon^T \text{ for all } \epsilon \in (0, \epsilon_u].$$

2. Assume that

$$uAu^{*T} = u^*Au^{*T} \text{ and } u \neq u^*.$$

Then it follows $u^*Au^T > uAu^T$ which implies

$$uAw_\epsilon^T < u^*Aw_\epsilon^T \iff w_\epsilon Aw_\epsilon^T < u^*Aw_\epsilon^T.$$

Result 1. *If $u^* \in \Delta$ is an evolutionarily stable Nash equilibrium, then, for every $u \in \Delta$ with $u \neq u^*$, there exists some $\epsilon_u \in (0, 1]$ such that*

$$w_\epsilon Aw_\epsilon^T < u^*Aw_\epsilon^T \text{ for all } \epsilon \in (0, \epsilon_u] \tag{2.6}$$

where

$$w_\epsilon = (1 - \epsilon)u^* + \epsilon u.$$

Conversely let $u^* \in \Delta$ be such that for every $u \in \Delta$ with $u \neq u^*$ there exists some $\epsilon_u \in (0, 1]$ such that (2.6) is satisfied. Then it follows from the equivalence (2.5) that

$$u^*Aw_\epsilon^T > uAw_\epsilon^T \tag{2.7}$$

and in turn for $\epsilon \to 0$ that

$$u^*Au^{*T} \geq uAu^*.$$

Now let $u^*Au^{*T} = uAu^{*T}$. Then it follows from (2.7) that

$$\begin{aligned}(1 - \epsilon)u^*Au^{*T} + \epsilon u^*Au^T &> (1 - \epsilon)uAu^{*T} + \epsilon uAu^T \\ &= (1 - \epsilon)u^*Au^{*T} + \epsilon uAu^T\end{aligned}$$

which implies $uAu^T < u^*Au^T$, i.e. u^* is an evolutionarily stable Nash equilibrium.

Result 2. *A population state $u^* \in \Delta$ is an evolutionarily stable Nash equilibrium, if and only if for every $u \in \Delta$ with $u \neq u^*$ there exists some $\epsilon_u \in (0, 1]$ such that the condition* (2.6) *is satisfied.*

From Corollary 1 it follows for an evolutionarily stable Nash equilibrium $u^* \in \Delta$ that

$$uAu^T < u^*Au^T \text{ for all } u \in \Delta \text{ with } u \neq u^* \text{ and } S(u) \subseteq S(u^*).$$

Now let $u \in \Delta$ be such that $S(u) \nsubseteq S(u^*)$. Then there exists some $i \in \{1, \ldots, n\}$ such that $u_i > 0$ and $u_i^* = 0$.

If $u_i \geq u_i^*$ for all $i = 1, \ldots, n$, then it follows from

$$\sum_{i=1}^{n} u_i = \sum_{i=1}^{n} u_i^* = 1$$

that $u = u^*$ which is impossible.

Hence there exists some $i \in \{1, \ldots, n\}$ with $u_i < u_i^*$. If we define

$$\lambda = \min \left\{ \frac{u_i^*}{u_i^* - u_i} | u_i < u_i^* \right\}$$

and put

$$v = u^* + \lambda(u - u^*),$$

then it follows that

$$v \in C = \left\{ u \in \Delta | \exists i_1 \text{ with } u_{i_1} > 0 \text{ and } u_{i_1}^* = 0 \text{ and } \exists i_2 \text{ with } u_{i_2} = 0 \right\}.$$

Conversely, if $v \in C$ is given and we define, for any $\lambda \in (0, 1]$, $u = u^* + \lambda(v - u^*)$, then $u \in \Delta$ and $S(u) \nsubseteq S(u^*)$.

By Result 1 we know that for every $v \in C$ there is some $\epsilon_v \in (0, 1]$ such that

$$w_\epsilon A w_\epsilon^T < u^* A w_\epsilon^T \text{ for all } \epsilon \in (0, \epsilon_v)$$

where

$$w_\epsilon = (1 - \epsilon)u^* + \epsilon v = u^* + \epsilon(v - u^*).$$

Since C is compact and $\epsilon_v, v \in C$, can be chosen continuously, there exists some $\hat{\epsilon} > 0$ with $\hat{\epsilon} = \min_{v \in C} \epsilon_v$ and therefore

$$w_\epsilon A w_\epsilon^T < u^* A w_\epsilon \text{ for all } \epsilon \in (0, \hat{\epsilon}].$$

If we define

$$\epsilon^* = \frac{\hat{\epsilon}}{\min_{v \in C} \|v - u^*\|_2},$$

then it follows that

$$u A u^T < u^* A u^T \text{ for all } u \in \Delta$$

with $S(u) \nsubseteq S(u^*)$ and $\|u - u^*\|_2 < \epsilon^*$. Summarising we obtain the

Theorem 2.2 *If $u^* \in \Delta$ is an evolutionarily stable Nash equilibrium, then there exists some $\epsilon^* > 0$ such that*

$$uAu^T < u^*Au^T \text{ for all } u \in \Delta \text{ with } u \neq u^* \text{ and } \|u - u^*\|_2 < \epsilon^* \tag{2.8}$$

which shows that u^* is the only Nash equilibrium in

$$V(u^*) = \{u \in \Delta | \; \|u - u^*\|_2 < \epsilon\}.$$

Conversely let $u^* \in \Delta$ and $\epsilon^* > 0$ be given such that the condition (2.8) is satisfied.

If we take any $u \in \Delta$ with $u \neq u^*$ and define, for $\epsilon \in (0, 1]$,

$$w_\epsilon = (1 - \epsilon)u^* + \epsilon u,$$

then $w_\epsilon \in \Delta,\; w_\epsilon \neq u^*$ and

$$\|w_\epsilon - u^*\|_2 = \epsilon\|u - u^*\| < \epsilon^*$$

for $\epsilon < \min\left(1, \frac{\epsilon}{\|u-u^*\|_2}\right) = \epsilon_u \in (0, 1)$ which implies

$$w_\epsilon A w_\epsilon^T < u^* A w_\epsilon^T$$

and shows that (2.6) is satisfied which implies by Result 2 that u^* is an evolutionarily stable Nash equilibrium.

Result 3. *A population state $u^* \in \Delta$ is an evolutionarily stable Nash equilibrium, if and only if there exists some $\epsilon^* > 0$ such that the condition* (2.8) *is satisfied.*

2.1.3 Evolutionarily Stable Equilibria for 2x2-Matrices

In this section we consider evolution-matrix-games with 2x2-payoff matrices:

$$A = \begin{pmatrix} a_{11} & a_{12} \\ a_{21} & a_{22} \end{pmatrix}.$$

We assume that

$$(a_{11} - a_{21})^2 + (a_{22} - a_{12})^2 > 0.$$

We distinguish three cases:

1. $a_{11} < a_{21}$ and $a_{22} < a_{12}$.

   Then we define
   $$u^* = \left(\frac{a_{12} - a_{22}}{d}, \frac{a_{21} - a_{11}}{d}\right) \tag{2.9}$$
   where
   $$d = (a_{12} - a_{22}) + (a_{21} - a_{11}) > 0.$$
   Then $u^* \in \Delta$ and $S(u^*) = \{1, 2\}$.

   Further it follows that
   $$e_1 A u^{*T} = e_2 A u^{*T} = \frac{a_{12}a_{21} - a_{11}a_{22}}{d} \text{ and}$$
   $$uAu^{*T} = u_1 e_1 A u^{*T} + u_2 e_2 A u^{*T} = u_1^* e_1 A u^{*T} + u_2^* e_2 A u^{*T} = u^* A u^{*T}$$
   for all $u^* \in \Delta$, i.e., u^* is a Nash equilibrium.

   Finally it follows that
   $$u^* A u^T - uAu^T = d(u_1^* - u_1)^2 > 0 \text{ for all } u \neq u^* (\iff u_1 \neq u_1^*),$$
   i.e., u^* is an evolutionarily stable Nash equilibrium.

   By Theorem 2.1 u^* is the only Nash equilibrium.

2. $a_{11} > a_{21}$ and $a_{22} > a_{12}$.

   Then it follows that
   $$uAe_1^T = a_{11}u_1 + a_{21}u_2 = a_{11} + \underbrace{(a_{21} - a_{11})}_{<0} u_2 < a_{11} = e_1 A e_1^T,$$
   for all $u \in \Delta$ with $u \neq e_1$ and
   $$uAe_2^T = a_{12}u_1 + a_{22}u_2 = a_{12} + a_{22} + \underbrace{(a_{12} - a_{22})}_{<0} u_1 < a_{22} = e_2 A e_2^T,$$
   for all $u \in \Delta$ with $u \neq e_2$.
   This implies that e_1 and e_2 are evolutionarily stable Nash equilibria.

3. $(a_{11} - a_{21})(a_{22} - a_{12}) < 0$.

   In this case either e_1 (if $a_{11} > a_{21}$ and $a_{22} < a_{12}$) or e_2 (if $a_{11} < a_{21}$ and $a_{22} > a_{12}$) is an evolutionarily stable Nash equilibrium.

Summarising we obtain the

Theorem 2.3 *If* $a_{11} \neq a_{21}$ *and* $a_{22} \neq a_{12}$*, then there exists at least one evolutionarily stable Nash equilibrium. In case 1) it is given by* (2.9) *and is the only Nash equilibrium, in case 2) it is* e_1 *or* e_2*, and in case 3) it is either* e_1 *or* e_2*.*

2.1.4 On the Detection of Evolutionarily Stable Equilibria

For $n > 3$ the existence of evolutionarily stable Nash equilibria cannot be guaranteed in general. For instance for the matrix

$$A = \begin{pmatrix} 1 & 2 & 0 \\ 0 & 1 & 2 \\ 2 & 0 & 1 \end{pmatrix}$$

there exist no evolutionarily stable Nash equilibria. In this case one can show that

$$uAu^T = 1 \text{ for all } u \in \Delta$$

and that

$$e_iAu^T \leq uAu^T = 1 \text{ is equivalent with } u_1 = u_2 = u_3 = \frac{1}{3}$$

which implies that $u^* = (\frac{1}{3}, \frac{1}{3}, \frac{1}{3})$ is a Nash equilibrium with

$$vAu^{*T} = u^*Au^{*T} \text{ for all } v \in \Delta.$$

But we also have

$$u^*Av^T = 1 = vAv^T \text{ for all } v \in \Delta$$

so that u^* cannot be evolutionarily stable. Since u^* is the only Nash equilibrium, no evolutionarily stable Nash equilibrium can exist.

We shall show later that the existence of Nash equilibria is ensured for every $n \geq 2$.

In the following we will present an algorithm developed in [1] by which all evolutionarily stable Nash equilibria can be detected or confirmed that none exists.

In order to describe this algorithm we need some preparations. So let NA be the set of all Nash equilibria (which is non-empty) and for every $u \in \Delta$ we define the set

$$J(u) = \left\{ i \in \{1, \ldots, n\} \,|\, e_iAu^T = uAu^T \right\}.$$

Then Lemma 1 can be rephrased in the form

$$u^* \in \mathrm{NA} \implies S(u^*) \subseteq J(u^*). \tag{2.10}$$

Let ESS denote the set of all evolutionarily stable Nash equilibria.

With these definitions we can prove the

Lemma 2. *If $u \in NA$ and $u^* \in ESS$ with $S(u) \subseteq J(u^*)$, then $u = u^*$.*

Proof. $S(u) \subseteq J(u^*)$ is equivalent to

$$u_i > 0 \implies e_i A u^{*T} = u^* A u^{*T}$$

which implies $uAu^{*T} = u^* A u^{*T}$.

Now let $u \neq u^*$, then it follows $u^* A u^T > uAu^T$, since $u^* \in \mathrm{ESS}$.

On the other hand, however, we have $u^* A u^T \leq uAu^T$, since $u \in \mathrm{NA}$. So $u \neq u^*$ is impossible. □

As a consequence of Lemma 2 and (2.10) we have the following statement: If $u^* \in \mathrm{ESS}$, then there is no Nash equilibrium $u \in \Delta$ with $S(u) \subseteq S(u^*)$ which is different from u^*. So the supports of evolutionarily stable Nash equilibria do not form chains with respect to set inclusion in the power set of $N = \{1, \ldots, n\}$.

This is the statement on which is based the following

Algorithm: We start with the

Initialization Step: At first we check whether a pure state e_i is a Nash equilibrium, i.e. whether

$$a_{ii} = \max_{j \in N} a_{ji}$$

holds true. Then we put

$$\mathrm{N}' = \mathrm{N} \setminus \{i \in \mathrm{N} | e_i \text{ is a Nash equilibrium }\}.$$

Now any mixed $u \in \mathrm{ESS}$ satisfies $S(u) \subseteq J(u) \subseteq \mathrm{N}'$ due to Lemma 2. Next we check whether a Nash equilibrium e_i is evolutionarily stable. If the answer is positive, we record e_i, put

$$\mathcal{S} = \{S \subseteq \mathrm{N}' | S \text{ has more than one element }\}$$

and proceed to the

Main Step: We denote by $\mathcal{S}_{\max}$ the system of all $S \in \mathcal{S}$ that are maximal with respect to set inclusion in $\mathcal{S}$. Then we check for any $S \in \mathcal{S}_{\max}$ whether there exists a Nash equilibrium $u \in \Delta$ with $S(u) = S$. If the answer is positive, we check whether $u \in$ ESS. If this is the case, we record u. If there are not any evolutionarily stable Nash equilibria with support in $\mathcal{S}_{\max}$, then we put

$$\mathcal{S}' = \mathcal{S} \setminus \mathcal{S}_{\max}.$$

Otherwise we denote the evolutionarily stable Nash equilibria with support in $\mathcal{S}_{\max}$ by $u^1, \ldots, u^s$, determine $J(u^i), i = 1, \ldots, s$ and put

$$\mathcal{S}' = \left\{S \in \mathcal{S} | S \not\subseteq J(u^i) \text{ for } i = 1, \ldots, s\right\} \setminus \mathcal{S}_{\max}.$$

If $\mathcal{S}'$ is empty, we stop, and the list of recorded evolutionarily stable Nash equilibria is complete.

Otherwise we denote by $\mathcal{S}_{\min}$ the system of all $S \in \mathcal{S}'$ which are minimal with respect to set inclusion in $\mathcal{S}'$. Then we check for any $S \in \mathcal{S}_{\min}$ whether there exists a Nash equilibrium $u \in \Delta$ with $S(u) = S$. If there is none such, then we put

$$\mathcal{S}'' = \mathcal{S}' \setminus \mathcal{S}_{\min}.$$

Otherwise we denote those Nash equilibria by $v^1, \ldots, v^r$ and check whether they are evolutionarily stable. If the answer is positive, the corresponding Nash equilibrium is recorded. Then we put

$$\mathcal{S}'' = \left\{S \in \mathcal{S}' | S(v^i) \not\subseteq S \text{ for } i = 1, \ldots, r\right\} \setminus \mathcal{S}_{\min}.$$

If $\mathcal{S}''$ is empty, we stop, and the list of recorded evolutionarily stable Nash equilibria recorded so far is complete.

Otherwise we repeat the main step with $\mathcal{S}''$ instead of $\mathcal{S}$.

The algorithm requires two subroutines: 1) For any non-empty subset $S \subseteq$ N it has to be decided whether there exists a Nash equilibrium $u \in \Delta$ with $S(u) \subseteq S$ or not.

2) For a given Nash equilibrium it has to be decided whether it is evolutionarily stable or not.

Here we will not go into the details of these subroutines and refer to [1]. Instead we will demonstrate the algorithm by an example which we take from [1].

Let $n = 5$ and

$$A = \begin{pmatrix} 1 & 0 & 2 & 2 & 2 \\ 0 & 1 & 2 & 2 & 2 \\ 2 & 2 & 1 & 0 & 0 \\ 2 & 2 & 0 & 1 & 0 \\ 2 & 2 & 0 & 0 & 1 \end{pmatrix}.$$

Then we have N'=N because of

$$a_{ii} < \max_{j \in \mathrm{N}} a_{ji} \text{ for all } i \in \mathrm{N}.$$

Therefore

$$\mathcal{S} = \{S \subseteq N | S \text{ has at least two elements }\}$$

and

$$\mathcal{S}_{\max} = \{\mathrm{N}\}.$$

By the Corollary 2 of Lemma 1 a population state $u \in \Delta$ with $S(u)$ =N is a Nash equilibrium, if and only if $J(u)$ =N, from which we infer that

$$u_1 = u_2 = a \text{ and } u_3 = u_4 = u_5 = b$$

where

$$2a + 3b = 1 \text{ and } a + 6b = 4a + b,$$

hence $a = \frac{5}{19}$, $b = \frac{3}{19}$. Thus $u = \left(\frac{5}{19}, \frac{5}{19}, \frac{3}{19}, \frac{3}{19}, \frac{3}{19}\right)$ is the only Nash equilibrium with $S(u)$ =N. It is, however, not evolutionarily stable, since for $v = \left(\frac{1}{2}, 0, 0, 0, \frac{1}{2}\right)$ it follows that

$$u^T A v^T = \frac{23}{19} < \frac{3}{2} = v A v^T.$$

Therefore

$$\mathcal{S}' = \mathcal{S} \setminus \mathcal{S}_{\max} = \{S \subseteq \mathrm{N} | S \text{ has at least two and at most four elements}\}.$$

Further we have

$$\mathcal{S}_{\min} = \{S \subseteq \mathrm{N} | S \text{ has exactly two elements}\}.$$

Now it is easy to see that no $u \in \Delta$ with

$$S(u) \in \{\{1, 2\}, \{3, 4\}, \{3, 5\}, \{4, 5\}\} \subseteq \mathcal{S}_{\min}$$

can be a Nash equilibrium. The remaining six possible supports of Nash equilibria are therefore of the form $\{i, k\}$ with $1 \leq i \leq 2$ and $3 \leq k \leq 5$.

We denote them by $S_1, \ldots, S_6$. Let us consider $u^1 \in \Delta$ with $S(u^1) = S_1 = \{1, 3\}$. Then u^1 is of the form $u^1 = (t, 0, 1-t, 0, 0)$ with $0 \le t \le 1$ and we get

$$Au^{1T} = (2-t, 2(1-t), t+1, 2t, 2t)^T.$$

From $e_i A u^{1T} = u^1 A u^{1T}$ for $i = 1, 3$, if u^1 is a Nash equilibrium, we conclude that this is the case, if and only if $t = \frac{1}{2}$, i.e.,

$$u^1 = \frac{1}{2}e_1 + \frac{1}{2}e_2 = \left(\frac{1}{2}, 0, \frac{1}{2}, 0, 0\right) \text{ and } u^1 A u^{1T} = \frac{3}{2}.$$

Now let $u = (t, 0, 1-t, 0, 0)$ with $0 \le t \le 1$ and $t \neq \frac{1}{2}$, then it follows that

$$u^1 A u^T = \frac{1}{2}(2-t) + \frac{1}{2}(t+1) = \frac{3}{2} \text{ and}$$
$$uAu^T = t(2-t) + (1-t)(t+1) = 1 + 2t(1-t) < \frac{3}{2}$$

which shows that u^1 is an evolutionarily stable Nash equilibrium. In a similar way one can show that every $u \in \Delta$ of the form $u = \frac{1}{2}e_i + \frac{1}{2}e_k$ for $1 \le i \le 2$ and $3 \le k \le 5$ belongs to ESS. Further we obtain

$\mathcal{S}'' = \{S \subseteq N$ with more than two elements $| S_j \not\subseteq S$ for all $j = 1, \ldots, 6\}$ which implies $\mathcal{S}'' = \{\{3, 4, 5\}\}$.

Since $\mathcal{S}'' \neq \emptyset$, we have to repeat the main step with $\mathcal{S}''$ instead of $\mathcal{S}$.

Because of $\mathcal{S}''_{\max} = \mathcal{S}''$ we have to check whether there exists a Nash equilibrium $u \in \Delta$ with $S(u) = \{3, 4, 5\}$. Every such u is of the form $u = (0, 0, u_3, u_4, u_5)$ and from $e_i A u^T = uAu^T$ for $i = 3, 4, 5$ it follows that $u_3 = u_4 = u_5 = \frac{1}{3}$.

This implies $uAu^T = \frac{1}{3} < 2 = e_i A u^T$ for $i = 1, 2$. Hence u is not a Nash equilibrium.

Thus $\mathcal{S}'' = \emptyset$ and the algorithm stops.

Result. *There exist exactly 6 elements in ESS which are given by* $\frac{1}{2}e_i + \frac{1}{2}e_k$, *for* $1 \le i \le 2$ *and* $3 \le k \le 5$.

2.1.5 A Dynamical Treatment of the Game

We start with an evolution-matrix-game as being introduced in Section 2.1.1.

We assume that

$$uAu^T > 0 \text{ for all } u \in \Delta.$$

We further assume that this game is submitted to a time discrete dynamics according to which the population states change as follows:

Let $u^k = (u_1^k, \ldots, u_n^k) \in \Delta$ be the population state in the k-th generation and let r_i^k be the average number of offsprings of individuals in the k-th generation that choose the strategy I_i. Then it follows for the next generation that

$$u_i^{k+1} = \frac{r_i^k u_i^k}{\sum_{j=1}^n r_j^k u_j^k}, \quad i = 1, \ldots, n.$$

Obviously, r_i^k depends on the average payoff to the I_i-individual which is given by $e_i A u^{k^T}$. We assume that

$$r_i^k = c e_i A u^{k^T}, \quad i = 1, \ldots, n,$$

where c is a positive constant. Then it follows that

$$u_i^{k+1} = \frac{c e_i A u^{k^T}}{c \sum_{j=1}^n e_j A u^{k^T} u_j^k} u_i^k = \frac{e_i A u^{k^T}}{u^k A u^{k^T}} u_i^k, i = 1, \ldots, n.$$

Obviously $u^k \in \Delta$ implies that $u^{k+1} \in \Delta$. Therefore, if we define a mapping $f_A : \Delta \to \Delta$ by

$$f_A(u)_i = \frac{e_i A u^T}{u A u^T} \quad \text{for } i = 1, \ldots, n \text{ and } u \in \Delta,$$

then $u^* \in \Delta$ is a fixed point of f_A, i.e.,

$$f_A(u^*) = u^*,$$

if and only if

$$e_i A u^{*^T} = u^* A u^{*^T} \text{ for all } i \in S(u^*). \tag{2.11}$$

By Lemma 1 this condition is necessary for u^* being a Nash equilibrium. This implies that $u^* \in \Delta$ is a fixed point of f_A, if u^* is a Nash equilibrium.

Conversely the question arises under which condition every fixed point of f_A is a Nash equilibrium. A first answer to this question is the

Lemma 3. *If* $u^* \in \Delta$ *is a fixed point of* f_A *and if*

$$u_i^* > 0 \textit{ for all } i = 1, \ldots, n, \tag{2.12}$$

then u^* *is a Nash equilibrium.*

Proof. $u^* \in \Delta$ is a fixed point of f_A, if and only if the condition (2.11) is satisfied. Because of $S(u^*) = \{1, \ldots, n\}$ this implies

$$uAu^{*T} = u^*Au^{*T} \text{ for all } u \in \Delta$$

which shows that u^* is a Nash equilibrium. □

A second answer to the above question is given by the

Theorem 2.4 *If* $u^* \in \Delta$ *is an attractive fixed point of* f_A, *i.e., a fixed point which is an attractor, then* u^* *is a Nash equilibrium.*

Proof. If $u^* \in \Delta$ satisfies (2.12), then the assertion follows from Lemma 3. If $S(u^*) \neq \{1, \ldots, n\}$, then (2.11) holds true. If we then show that

$$e_iAu^{*T} \leq u^*Au^{*T} \text{ for all } i \in \{1, \ldots, n\} \setminus S(u^*),$$

it follows that u^* is a Nash equilibrium. □

Let us assume that there is a $k \in \{1, \ldots, n\} \setminus S(u^*)$ such that

$$e_kAu^{*T} > u^*Au^{*T}. \tag{2.13}$$

Since $g(u) = e_kAu^T - uAu^T$ is continuous there is an $\epsilon_1 > 0$ such that

$$e_kAu^T > uAu^T \text{ for all } u \in \Delta \text{ with } \|u - u^*\|_2 < \epsilon_1. \tag{2.14}$$

Since u^* is an attractor, there exists an $\epsilon_2 > 0$ such that

$$\lim_{t\to\infty} f_A^t(u) = u^* \text{ for all } u \in \Delta \text{ with } \|u - u^*\|_2 < \epsilon_2. \tag{2.15}$$

This implies that for every $v \in \Delta$ with $\|v - u^*\|_2 < \epsilon$ there exists some $T_\epsilon \in \mathbb{N}$ such that

$$\|v(t) - u^*\|_2 < \epsilon \text{ for all } t \geq T_\epsilon$$

where $\epsilon = \min(\epsilon_1, \epsilon_2)$ and $v(t) = f_A^t(v)$.

From (2.13) we deduce that

$$v_k(t+1) > v_k(t) \text{ for all } t \geq T_\epsilon. \tag{2.16}$$

On the other hand it follows from (2.15) that

$$\lim_{t\to\infty} v_k(t) = u_k^* = 0, \quad \text{since } k \notin S(u^*)$$

which is a contradiction to (2.16). Therefore the assumption (2.13) is false and Theorem 2.3 is proved.

Next we investigate the question under which conditions a Nash equilibrium is an attractive fixed point of f_A.

At first we consider a Nash equilibrium $u^* \in \Delta$ with $S(u^*) = \{1, \ldots, n\}$. Then it follows from Corollary 2 of Lemma 1 that u^* is a fixed point of f_A. In general u^* need not be an attractive fixed point (see [4]).
We can, however, prove the following

Theorem 2.5 *Let $u^* \in \Delta$ be an evolutionarily stable Nash equilibrium with $S(u^*) = \{1, \ldots, n\}$. Further let K, L be two non-empty subsets of $\{1, \ldots, n\}$ with $K \cap L = \emptyset$ and $K \cup L = \{1, \ldots, n\}$ such that*

$$e_i A u^T > u A u^T \textit{ for all } u \in \Delta \textit{ with } 0 < u_i < u_i^* \textit{ and } i \in K$$

and

$$e_i A u^T < u A u^T \textit{ for all } u \in \Delta \textit{ with } u_i^* < u_i < 1 \textit{ and } i \in L.$$

Then for every

$$u^\circ \in U = \{u \in \Delta | 0 < u_i < u_i^* \textit{ for all } i \in K \textit{ and } u_i^* < u_i < 1 \textit{ for all } i \in L\}$$

it follows that

$$\lim_{k\to\infty} f_A^k(u^\circ) = u^*, \textit{ if } f_A(U) \subseteq U.$$

Proof. At first we have

$$0 < u_i < f_A(u)_i < u_i^* \text{ for all } u \in \Delta \text{ with } 0 < u_i < u_i^* \text{ for } i \in K$$

and

$$u_i^* < f_A(u)_i < u_i < 1 \text{ for all } u \in \Delta \text{ with } u_i^* < u_i < 1 \text{ for all } i \in L.$$

If we choose $u^\circ \in U$ arbitrarily and define

$$u^k = f_A^k(u^\circ) \text{ for } k \in \mathbb{N}_0,$$

then it follows that $u^k \in U$ for all $k \in \mathbb{N}_0$ and because of $u^{k+1} = f_A(u^k)$, $k \in \mathbb{N}_0$, we infer that $u^k \to \hat{u} = \overline{U}$ with $S(\hat{u}) = \{1, \ldots, n\}$ and $\hat{u} = f_A(\hat{u})$. By Lemma 3 $\hat{u}$ is also a Nash equilibrium. According to Theorem 2.1 u^* is the only Nash equilibrium which implies that $\hat{u} = u^*$.

This completes the proof. □

Let us consider the case $n = 2$ where $A = \begin{pmatrix} a_{11} & a_{12} \\ a_{21} & a_{22} \end{pmatrix}$.
We assume $a_{11} < a_{21}$ and $a_{22} < a_{12}$.

In Section 2.1.3 we have shown that

$$u^* = \left(\frac{a_{12} - a_{22}}{d}, \frac{a_{21} - a_{11}}{d} \right)$$

with

$$d = (a_{12} - a_{22}) + (a_{21} - a_{11})$$

is an evolutionarily stable Nash equilibrium which by Theorem 2.1 is the only Nash equilibrium.

Further one can show that, for all $u \in \Delta$,

$$e_1 A u^T - u A u^T = d(u_1 - 1)(u_1 - u_1^*)$$

and

$$e_2 A u^T - u A u^T = d u_1 (u_1 - u_1^*) = d(1 - u_2)(u_2^* - u_2)$$

which implies

$$e_1 A u^T - u A u^T > 0 \text{ for all } u \in \Delta \text{ with } 0 < u_1 < u_1^*$$

and

$$e_2 A u^T - u A u^T < 0 \text{ for all } u \in \Delta \text{ with } u_2^* < u_2 < 1.$$

If we define

$$U = \{u \in \Delta \mid 0 < u_1 < u_1^*\} = \{u \in \Delta \mid u_2^* < u_2 < 1\}$$

and assume that $f_A(U) \subseteq U$, then all the assumptions of Theorem 2.5 are satisfied with $K = \{1\}$ and $L = \{2\}$ which implies

$$\lim_{k\to\infty} f_A^k(u^\circ) = u^* \quad \text{for all } u^\circ \in U.$$

Example:

$A = \begin{pmatrix} 0 & 2 \\ 1 & 1 \end{pmatrix}$; then $a_{11} = 0 < a_{21} = 1$ and $a_{22} = 1 < a_{12} = 2$.
Then $u^* = (\frac{1}{2}, \frac{1}{2})$ and for every $u \in \Delta$ it follows that

$$f_A(u)_1 - \frac{1}{2} = \frac{2u_1}{2u_1 + 1} - \frac{2u_1 + 1}{2(2u_1 + 1)} = \frac{2u_1 - 1}{2(2u_1 + 1} < 0, \text{ if } 0 < u_1 < \frac{1}{2},$$

hence,

$$f_A(u)_1 < \frac{1}{2} \Rightarrow f_A(U) \subseteq U.$$

For pure population states one can prove the following

Theorem 2.6 *If, for some $k \in \{1, \ldots, n\}$, e_k is an evolutionarily stable Nash equilibrium, then e_k is an asymptotically stable fixed point of f_A.*

Proof. By Theorem 2.2 there exists some $\epsilon^* > 0$ such that

$$uAu^T < e_kAu^T \text{ for all } u \in \Delta \text{ with } u \neq e_k \text{ and } \|u - e_k\|_2 < \epsilon^*.$$

Let us define

$$U = \{u \in \Delta | \; \|u - e_k\|_2 < \epsilon^*\}.$$

Then it follows that

$$f_A(u)_k = \frac{e_kAu^T}{uAu^T} u_k \geq u_k \text{ for all } u \in U.$$

If we define a continuous function $V : \Delta \to \mathbb{R}$ by

$$V(u) = 1 - u_k, \; u \in \Delta,$$

then it follows

$$V(f_A(u)) - V(u) = u_k - f_A(u)_k \leq 0 \text{ for all } u \in U.$$

This shows that V is a Lyapunov function with respect to f and U.

Further we infer that

$$V(u) \geq 0 \text{ for all } u \in U \text{ and } (V(u) = 0 \iff u = e_k),$$

i.e., V is positive definite with respect to e_k.

Finally we obtain

$$V(f_A(u)) - V(u) < 0 \text{ for all } u \in U \text{ with } u \neq e_k.$$

Thus all the assumptions of Satz 5.8 in [3] (see also Section A.3) are satisfied which implies that e_k is an asymptotically stable fixed point of f_A. □

Corollary. *Let, for some $k \in \{1, \ldots, n\}$,*

$$\begin{aligned} & a_{kk} \geq a_{jk} \textit{ for all } j = 1, \ldots, n \\ \textit{and} \quad & \\ & a_{kk} = a_{jk} \implies a_{ki} > a_{ji} \textit{ for all } i \neq k. \end{aligned} \tag{2.17}$$

Then e_k is an asymptotically stable fixed point of f_A.

Proof. In Section 2.1.1 we have shown that the conditions (2.17) imply that e_k is an evolutionarily stable Nash equilibrium. Thus the assertion follows from Theorem 2.6. □

2.1.6 Existence and Iterative Calculation of Nash Equilibria

We start with a necessary and sufficient condition for a Nash equilibrium $u^* \in \Delta$ which is given by

$$u^* A u^{*T} = \max_{i=1,\ldots,n} e_i A u^{*T}. \tag{2.18}$$

Now we define, for every $u \in \Delta$ and every $i = 1, \ldots, n$,

$$\varphi_i(u) = \max(0, e_i A u^T - u A u^T). \tag{2.19}$$

From (2.18) we then deduce that $u^* \in \Delta$ is a Nash equilibrium, if and only if

$$\varphi_i(u^*) = 0 \text{ for } i = 1, \ldots, n. \tag{2.20}$$

With the aid of the functions (2.19) we now define a mapping $f : \Delta \to \Delta$ via

$$f_i(u) = \frac{1}{1 + \sum_{j=1}^{n} \varphi_j(u)} (u_i + \varphi_i(u)), \; i = 1, \ldots, n, \; u \in \Delta.$$

This mapping is continuous and by Brouwer's fixed point theorem it has a fixed point $u^* \in \Delta$, i.e., $f(u^*) = u^*$.

Assertion. *The condition* (2.20) *is satisfied, i.e., u^* is a Nash equilibrium.*

Proof. We choose $i_0 \in \{1, \ldots, n\}$ such that $e_{i_0} A u^{*T} = \min_{u_j^* > 0} e_j A u^{*T}$ and infer that

$$e_{i_0} A u^{*T} \leq u^* A u^{*T} = \sum_{j=1}^{n} u_j^* e_j A u^{*T}$$

which implies $\varphi_{i_0}(u^*) = 0$.

This implies further that

$$u_{i_0}^* = \frac{u_{i_0}^*}{1 + \sum_{j=1}^{n} \varphi_j(u^*)}, \text{ hence } \sum_{j=1}^{n} \varphi_j(u^*) = 0 \implies (2.20).$$

□

Result. *Every fixed point of f is a Nash equilibrium.*

Conversely, every Nash equilibrium is a fixed point of f, since (2.20) implies $f(u^*) = u^*$. For the calculation of fixed points it is natural to perform an iteration method which starts with an initial state $u^\circ \in \Delta$ and creates a sequence of states $u^k \in \Delta, k \in \mathbb{N}_0$, according to the recursion $u^{k+1} = f(u^k), k \in \mathbb{N}_0$. If this sequence converges to some $u^* \in \Delta$, then u^* is a fixed point of f and hence a Nash equilibrium.

The question now arises under which conditions the sequence $(u^{k+1} = f(u^k))$ with $u^\circ \in \Delta$ converges to some $u^* \in \Delta$. In order to give an answer to this question we assume that there exists a subset $U \subseteq \Delta$ such that

$$\varphi_j(u) = 0 \text{ for all } j \in J \subseteq \{1, \ldots, n\}, \; |J| < n, \text{ and all } u \in U.$$

Then it follows, for every $u \in U$, that

$$f_i(u) = \frac{u_i}{1 + \sum\limits_{j \in J} \varphi_j(u)}, \text{ for } i \in J$$

and

$$f_i(u) = \frac{u_i + \varphi_i(u)}{1 + \sum\limits_{j \in J} \varphi(u_j)}, \text{ for } i \notin J.$$

This implies, for every $u \in U$, that

$$f_i(u) \leq u_i \text{ for all } i \in J. \tag{2.21}$$

Assumption: $f(U) \subseteq U$.

If we define, starting with some $u^\circ \in U$, a sequence $(u^k)_{k \in \mathbb{N}_0}$ in U by $u^{k+1} = f(u^k), k \in \mathbb{N}_0$, then it follows from (2.21) that for every $i \in J$ there exists some $u_i^* \in [0, 1]$ with $u_i^* = \lim\limits_{k \to \infty} u_i^k$.

Let us assume that also for every $i \notin J$ the sequence $(u_i^k)_{k \in \mathbb{N}_0}$ converges to some $u_i^* \in [0, 1]$ so that the sequence $(u^k)_{k \in \mathbb{N}_0}$ converges to some $u^* \in \overline{U} \subseteq \Delta$ which is a fixed point and hence a Nash equilibrium.

In the special case $J = \{1, \ldots, n\} \setminus \{i_0\}$ for some $i_0 \in \{1, \ldots, n\}$ we infer from (2.21) that

$$f_{i_0}(u) = 1 - \sum_{j \in J} f_i(u) \geq 1 - \sum_{j \in J} u_i = u_{i_0}$$

which implies that the sequence $(u_{i_0}^k)_{k \in \mathbb{N}_0}$ converges to some $u_{i_0}^* \in [0, 1]$.

Further we have

$$\begin{aligned} e_{i_0} A u^{*T} - u^* A u^{*T} &= \sum_{j=1}^{n} a_{i_0, j} u_j^* - \sum_{i=1}^{n} u_i^* \sum_{j=1}^{n} a_{ij} u_j^* \\ &= \sum_{\substack{i=1 \\ i \neq i_0}}^{n} u_i^* \sum_{j=1}^{n} (a_{i_0 j} - a_{ij}) u_j^* \leq 0. \end{aligned}$$

Assumption:

$$a_{i_0 j} - a_{ij} > 0 \text{ for all } i \neq i_0 \text{ and } j = 1, \ldots, n. \tag{2.22}$$

Then it follows that

$$u_i^* = 0 \text{ for all } i \neq i_0, \text{ hence, } u_{i_0}^* = 1 \text{ and thus } u^* = e_{i_0}.$$

We will demonstrate this by an example: Let $n = 3$ and

$$A = \begin{pmatrix} 9 & 6 & 3 \\ 8 & 5 & 2 \\ 7 & 4 & 1 \end{pmatrix}.$$

Then we obtain

$$\begin{aligned} e_1 A u^T &= 9u_1 + 6u_2 + 3u_3, \\ e_2 A u^T &= 8u_1 + 5u_2 + 2u_3, \\ e_3 A u^T &= 7u_1 + 4u_2 + u_3, \\ u A u^T &= 9u_1 + 6u_2 + 3u_3 - u_2 - 2u_3. \end{aligned}$$

This implies

$$\begin{aligned} e_1 A u^T - u A u^T &= u_2 + 2u_3 > 0 \\ &\quad \text{for all } \; u \in \Delta \text{ with } u_2 + 2u_3 > 0 \iff u_1 - u_3 < 1, \\ e_2 A u^T - u A u^T &= -1 + u_2 + 2u_3 < 0 \\ &\quad \text{for all } \; u \in \Delta \text{ with } u_2 + 2u_3 < 1 \iff u_3 < u_1, \\ e_3 A u^T - u A u^T &= -2 + u_2 + 2u_3 < 0 \\ &\quad \text{for all } \; u \in \Delta \text{ with } u_2 + 2u_3 < 2 \iff 0 < u_2 + 2u_1, \end{aligned}$$

and in turn

$$\varphi_1(u) > 0 \text{ for all } u \in \Delta = \{u \in U | 0 < u_3 < u_1 < 1\}$$

and

$$\varphi_2(u) = \varphi_3(u) = 0 \text{ for all } u \in U.$$

As a consequence we obtain

$$f_i(u) < u_i \text{ for } i = 2, 3 \text{ and } f_1(u) > u_1$$

for all $u \in U$ which implies (2.21) with $J = \{2, 3\}$. We also obtain $f(U) \subseteq U$ and the assumption (2.22) is satisfied.

In the general case $J \subseteq \{1, \ldots, n\} \setminus \{i_0\}$ it follows from the assumption (2.22) that e_{i_0} which we have shown to be a Nash equilibrium is evolutionarily stable.

In particular it follows from (2.22) that

$$a_{i_0 i_0} > a_{i i_0} \text{ for all } i \neq i_0$$

and hence

$$uAe_{i_0}^T = \sum_{i=1}^{n} u_i a_{i i_0} < a_{i_0 i_0} = e_{i_0} A e_{i_0}^T \text{ for all } u \in \Delta \text{ with } u \neq e_{i_0}.$$

This implies that $uAe_{i_0}^T = e_{i_0}Ae_{i_0}^T$ is only possible for $u = e_{i_0}$ which shows that e_{i_0} is evolutionarily stable.

Now let e_{i_0} be an evolutionarily stable Nash equilibrium. By Theorem 2.2 then there exists a relatively open subset $U \subseteq \Delta$ with $e_{i_0} \in U$ and

$$uAu^T < e_{i_0}Au^T \text{ for all } u \in U \text{ with } u \neq e_{i_0}.$$

This implies that

$$\varphi_{i_0}(u) > 0 \text{ for all } u \in U \text{ with } u \neq e_{i_0}.$$

Assumption:

$$\varphi_i(u) = 0 \text{ for all } u \in U \text{ and } i \neq i_0.$$

Then it follows that

$$f_i(u) < u_i \text{ for all } u \in U \text{ with } i \neq i_0$$

and

$$f_{i_0}(u) > u_{i_0} \text{ for all } u \in U \text{ with } u \neq e_{i_0}.$$

If we define a Lyapunov function $V : \Delta \to \mathbb{R}$ by

$$V(u) = 1 - u_{i_0}, u \in \Delta.$$

then it follows that

$$V(e_{i_0}) = 0 \text{ and } V(u) > 0 \text{ for all } u \in \Delta \text{ with } u \neq e_{i_0}$$

and

$$V(f(u)) - V(u) = 1 - f_{i_0}(u) - 1 + u_{i_0} < 0 \text{ for } u \in \Delta \text{ with } u \neq e_{i_0}.$$

By Satz 5.8 in [3] (see also Section A.3) e_{i_0} is an asymptotically stable fixed point of f.

This is in particular the case, if the assumption (2.22) is satisfied and the above assumption holds true.

Next we investigate the question under which conditions a Nash equilibrium $u^* \in \Delta$ is an asymptotically stable fixed point of f which in particular implies that there is a relatively open neighbourhood $W \subseteq \Delta$ of u^* such that for every $u^\circ \in W$ the sequence $(u^{k+1} = f(u_k))_{k \in \mathbb{N}_0}$ converges to u^*.

For this purpose we assume that there is a non-empty subset J of $\{1, \ldots, n\}$ such that

$$e_i A u^{*T} < u^* A u^{*T} \text{ for all } i \in J.$$

By Lemma 1 it then follows that

$$u_i^* = 0 \text{ for all } i \in J.$$

Further there exists a relatively open subset $U \subseteq \Delta$ with $u^* \in U$ such that

$$e_i A u^T < u A u^T \text{ for all } i \in J \text{ and all } u \in U.$$

This implies

$$\varphi_i(u) = 0 \text{ for all } i \in J \text{ and all } u \in U$$

and hence

$$f_i(u) = \frac{u_i}{1 + \sum_{j \notin J} \varphi_j(u)} \text{ for all } i \in J \text{ and all } u \in U.$$

Assumption: For every $u \in U$ there is some $i \in J$ with $u_i > 0$ and for every $u \in U$ with $u \neq u^*$ there is some $j \notin J$ with $\varphi_j(u) > 0$.

Then it follows that

$$f_i(u) \leq u_i \text{ for all } i \in J \text{ and all } u \in U \text{ with } u \neq u^*$$

and

$$f_i(u) < u_i \text{ for some } i = i(u) \in J.$$

If we define a Lyapunov function $V : U \to \mathbb{R}$ by

$$V(u) = \sum_{i \in J} u_i, \; u \in U,$$

then it follows that

$$V(u^*) = 0 \text{ and } V(u) > 0 \text{ for all } u \in U \text{ with } u \neq u^*$$

and

$$V(f(u)) = \sum_{i \in J} f_i(u) < \sum_{i \in J} u_i = V(u) \text{ for all } u \in U \text{ with } u \neq u^*.$$

Satz 5.8 in [3] (see also Section A.3) then implies that u^* is an asymptotically stable fixed point of f.

Finally we consider the case $n = 2$ with

$$A = \begin{pmatrix} a_{11} & a_{12} \\ a_{21} & a_{22} \end{pmatrix} \text{ and } a_{11} < a_{21}, a_{22} < a_{12}.$$

In Section 2.1.3 we have shown that

$$u^* = \left(\frac{a_{12} - a_{22}}{d}, \frac{a_{21} - a_{11}}{d} \right)$$

with

$$d = (a_{12} - a_{22}) - (a_{21} - a_{11})$$

is an evolutionarily stable Nash equilibrium.

Further we get

$$e_1 A u^T - u A u^T = d(u_1 - 1)(u_1 - u_1^*)$$

and

$$e_2 A u^T - u A u^T = d u_1 (u_1 - u_1^*) = d(1 - u_2)(u_2^* - u_2)$$

for all $u \in \Delta$.

If we put

$$U = \{u \in \Delta \mid u_1^* < u_1\},$$

then it follows that

$$\varphi_1(u) = 0 \text{ and } \varphi_2(u) > 0 \text{ for all } u \in U$$

which implies

$$f_1(u) = \frac{u_1}{1 + \varphi_2(u)} < u_1 \text{ and } f_2(u) = \frac{u_2 + \varphi_2(u)}{1 + \varphi_2(u)} = 1 - f_1(u) > 1 - u_1 = u_2$$

for all $u \in U$. Further we obtain

$$f_1(u) - u_1^* = \frac{(u_1 - u_1^*)(1 - du_1 u_1^*)}{1 + du_1(u_1 - u_1^*)} > \frac{(u_1 - u_1^*)(1 - du_1^*)}{1 + du_1(u_1 - u_1^*)} \geq 0$$

for all $u \in U$, if $1 - du_1^* = 1 - a_{12} + a_{22} \geq 0$.
This implies $f(U) \subseteq U$, if $1 - a_{12} + a_{22} \geq 0$. If we choose $u^\circ \in U$ and define a sequence $(u^k)_{k \in \mathbb{N}_0}$ in U by $u^{k+1} = f(u^k), k \in \mathbb{N}_0$, then it follows that $\lim_{k \to \infty} u^k = \hat{u} \in \overline{U}$ with $f(\hat{u}) = \hat{u}$. Since $\hat{u}$ is a Nash equilibrium, it follows by Theorem 2.1 that $\hat{u} = u^*$.
An example:
$A = \begin{pmatrix} 0 & 1 \\ 1 & 0 \end{pmatrix}$. Then $u^* = (\frac{1}{2}, \frac{1}{2})$ and $1 - a_{12} + a_{22} = 0$.
Let us end this section with a direct method for calculating Nash equilibria. We start with the statement that $\hat{u} \in \Delta$ is a Nash equilibrium, if and only if

$$\hat{u} A \hat{u}^T = \max_{i=1,\ldots,n} e_i A \hat{u}^T \tag{2.23}$$

Assertion. *$\hat{u} \in \Delta$ is a Nash equilibrium, if and only if*

$$\hat{u}_k > 0 \Rightarrow e_k A \hat{u}^T = \max_{i=1,\ldots,n} e_i A \hat{u}^T. \tag{2.24}$$

Proof. 1. Let (2.23) hold true. Then it follows that

$$\hat{u} A \hat{u}^T = \sum_{\substack{k=1 \\ \hat{u}_k > 0}}^{n} \hat{u}_k (A \hat{u}^T)_k = \sum_{\substack{k=1 \\ \hat{u}_k > 0}}^{n} \hat{u}_k (e_k \hat{u}^T)_k = \max_{i=1,\ldots,n} e_i A \hat{u}^T$$

i.e., (2.24) is true which implies that $\hat{u} \in \Delta$ is a Nash equilibrium.

2. Conversely let $\hat{u} \in \Delta$ is a Nash equilibrium. Then (2.23) holds true which implies

$$\max_{i=1,\ldots,n} e_i A \hat{u}^T = \hat{u} A \hat{u}^T = \sum_{k=1}^{n} \hat{u}_k (e_k A \hat{u}^T = \sum_{\substack{k=1 \\ \hat{u}_k > 0}}^{n} \hat{u}_k (e_k A \hat{u}^T).$$

From this it follows that (2.24) must be true.

From (2.23) and (2.24) we conclude that $\hat{u} \in \Delta$ is a Nash equilibrium, if and only if

$$(\hat{u}A\hat{u}^T)e \geq A\hat{u}^T$$

and

$$\hat{u}(A\hat{u}^T - (\hat{u}A\hat{u}^T)e^T) = 0 \quad \text{where } e = (1, \ldots, 1).$$

If we put

$$\hat{v}^T = -A\hat{u}^T + (\hat{u}A\hat{u}^T)e^T,$$

then the last two conditions are equivalent to

$$\hat{v}^T \leq \Theta_n = \text{ zero vector of } \mathbb{R}^n \text{ and } \hat{u}\hat{v}^T = 0.$$

These two conditions are therefore necessary and sufficient for $\hat{u} \in \Delta$ to be a Nash equilibrium. □

Now let $u^T \in \mathbb{R}^n_+$ with $u^T \neq \Theta_n$ be given such that

$$v^T = -Au^T + e^T \geq \Theta_n \quad \text{and} \quad uv^T = 0. \tag{2.25}$$

Then it follows that

$$uAu^T = ue^T > 0.$$

If we put

$$\hat{u} = \frac{1}{ue^T}u, \tag{2.26}$$

then we obtain

$$\hat{u} \in \Delta \quad \text{and} \quad \hat{u}A\hat{u} = \frac{1}{(ue^T)^2}uAu^T = \frac{1}{ue^T}.$$

Further we get

$$-A\hat{u}^T = -\frac{1}{ue^T}Au^T \geq -\frac{1}{ue^T}e^T = -(\hat{u}A\hat{u}^T)e^T$$

which implies

$$\hat{v}^T = -A\hat{u}^T + (\hat{u}A\hat{u}^T)e^T \geq \Theta_n$$

and

$$\hat{u}\hat{v}^T = \frac{1}{ue^T}u\left(-\frac{1}{ue^T}Au^T + \frac{1}{ue^T}e^T\right) = \frac{1}{(ue^T)^2}uv^T = 0.$$

Therefore $\hat{u}$ is a Nash equilibrium.

In order to find a Nash equilibrium $\hat{u} \in \varDelta$ we have to find a solution $u^T \in \mathbb{R}^n_+$ of (2.25) with $u^T \neq \Theta_n$ and to define $\hat{u}$ by (2.26).

Now let $S \subseteq N = \{1, \ldots, n\}$ be a non-empty subset.
Then we define

$$A_S = (a_{ij})_{i,j \in S}, e_S = (1, \ldots, 1)^T \in \mathbb{R}^{|S|}, A_{N,S} = (a_{ij})_{i \in N, j \in S}.$$

We assume that A_S is non-singular. Then it follows, for a solution $u^T \in \mathbb{R}^n_+$ of (2.25) with

$$u_i > 0 \quad \text{for } i \in S \quad \text{and} \quad u_i = 0 \quad \text{for } i \notin S, \tag{2.27}$$

that, for $u_S = (u_i)_{i \in S}$,

$$A_S u_S = e_S \quad \text{and} \quad A_{N,S} u_S \leq e_N = (1, \ldots, 1)^T \in \mathbb{R}^n$$

which implies

$$(u_S =)\, A_S^{-1} e_S > \Theta_{|S|} \quad \text{and} \quad A_{N,S} A_S^{-1} e_S \leq e_N. \tag{2.28}$$

If conversely the conditions (2.28) are satisfied, then $u^T \in \mathbb{R}^n$ with

$$u_S = A_S^{-1} e_s > \Theta_{|S|} \quad \text{and} \quad u_{N \setminus S} = \Theta_{|N \setminus S|}$$

is a solution of (2.25) with $u^T \neq \Theta_n$.

In particular for $S = \{i_0\}$ the conditions (2.28) read

$$a_{i_0 i_0} > 0 \quad \text{and} \quad a_{j i_0} \leq a_{i_0 i_0} \quad \text{for all } j \in N. \tag{2.29}$$

Examples:

1. $A = \begin{pmatrix} 2 & 4 \\ 1 & 3 \end{pmatrix} = B^T$. Here (2.29) is satisfied for $i_0 = 1$ and $u = (\frac{1}{2}, 0)^T$ is a solution of (2.25) with $u^T \neq \Theta_2$.
2. $A = \begin{pmatrix} 1 & 4 \\ 2 & 3 \end{pmatrix} = B^T$. Here (2.29) is not satisfied for any $i \in \{1, 2\}$.

For $S = \{1, 2\}$, however, we have

$$A_S^{-1} e_S = \begin{pmatrix} \frac{1}{5} \\ \frac{1}{5} \end{pmatrix} > \Theta_2$$

and

$$A_{N,S} A_S^{-1} e_S = \Theta_2,$$

i.e., the conditions (2.28) are satisfied and $u = (\frac{1}{5}, \frac{1}{5})$ is a solution of (2.25) with $u^T \neq \Theta_2$.

The inequality system in (2.25) can be represented by the following tableau:

		$-u_1$	$-u_2$	$\dots$	$-u_n$
v_1	1	a_{11}	a_{12}	$\dots$	a_{1n}
v_2	1	a_{21}	a_{22}	$\dots$	a_{2n}
$\vdots$	$\vdots$	$\vdots$	$\vdots$	$\ddots$	$\vdots$
v_n	1	a_{n1}	a_{n2}	$\dots$	a_{nn}

In order to obtain a solution of (2.25) with (2.27) one has to exchange with the aid of a Jordan elimination step u_i and v_i for every $i \in S$. The positive components of the solution are then in the first column of the tableau.
Let us demonstrate this by the following example: Let $n = 3$ and

$$A = \begin{pmatrix} 7 & 5 & 3 \\ 8 & 4 & 2 \\ 9 & 5 & 1 \end{pmatrix} = B^T.$$

Then the beginning tableau reads:

		$-u_1$	$-u_2$	$-u_3$
v_1	1	7	5	3
v_2	1	8	4	2
v_3	1	9	5	1

In order to obtain a solution of (2.25) with (2.27) for $S = \{1, 3\}$ one has to exchange u_1 with v_1 and u_3 with v_3. This leads to the tableau:

		$-v_1$	$-u_2$	$-v_3$
u_1	$\frac{1}{10}$	$-\frac{1}{20}$	$\frac{1}{2}$	$\frac{3}{20}$
v_2	0	$\frac{1}{2}$	-1	$\frac{1}{2}$
u_3	$\frac{1}{10}$	$-\frac{9}{20}$	$\frac{1}{2}$	$-\frac{7}{20}$

The solution therefore reads $u = (\frac{1}{10}, 0, \frac{1}{10})^T$.

This procedure can also be used in order to find out, for a given non-empty set $S \subseteq N$, whether there exists a solution of (2.25) with (2.27) for S or not.

We demonstrate this by the following example: Let $n = 5$ and

$$A = \begin{pmatrix} 1\,0\,2\,2\,2 \\ 0\,1\,2\,2\,2 \\ 2\,2\,1\,0\,0 \\ 2\,2\,0\,1\,0 \\ 2\,2\,0\,0\,1 \end{pmatrix} = B^T .$$

The beginning tableau reads:

		$-u_1$	$-u_2$	$-u_3$	$-u_4$	$-u_5$
v_1	1	1	0	2	2	2
v_2	1	0	1	2	2	2
v_3	1	2	2	1	0	0
v_4	1	2	2	0	1	0
v_5	1	2	2	0	0	1

The condition (2.29) is for no $i_0 \in \{1, 2, 3, 4, 5\}$ satisfied. Therefore there is no solution of (2.25) with (2.27) for $S = \{i\}, i \in N$.

For $S = \{1, 2\}$ one obtains by exchange of u_1 with v_1 and u_2 with v_2 the tableau

		$-v_1$	$-v_2$	$-u_3$	$-u_4$	$-u_5$
u_1	1	1	0	2	2	2
u_2	1	0	1	2	2	2
v_3	−3	−2	−2	−7	−8	−8
v_4	−3	−2	−2	−8	−7	−8
v_5	−3	−2	−2	−8	−8	−7

which shows that there is no solution of (2.25) with (2.27) for $S = \{1, 2\}$.
For $S = \{1, 3\}$ one obtains by exchange of u_1 with v_1 and u_3 with v_3 the tableau

		$-v_1$	$-u_2$	$-v_3$	$-u_4$	$-u_5$
u_1	$\frac{1}{3}$	$-\frac{1}{3}$	$\frac{4}{3}$	$\frac{2}{3}$	$\frac{2}{3}$	$\frac{2}{3}$
v_2	$\frac{1}{3}$	$-\frac{4}{3}$	$\frac{7}{3}$	$\frac{2}{3}$	$\frac{2}{3}$	$\frac{2}{3}$
u_3	$\frac{1}{3}$	$\frac{2}{3}$	$-\frac{2}{3}$	$-\frac{1}{3}$	$\frac{4}{3}$	$\frac{4}{3}$
v_4	$\frac{1}{3}$	$\frac{2}{3}$	$-\frac{2}{3}$	$-\frac{4}{3}$	$-\frac{7}{3}$	$\frac{4}{3}$
v_5	$\frac{1}{3}$	$\frac{2}{3}$	$-\frac{2}{3}$	$-\frac{4}{3}$	$\frac{4}{3}$	$\frac{7}{3}$

The solution of (2.25) therefore reads $u = (\frac{1}{3}, 0, \frac{1}{3}, 0, 0)^T$.

In the same way one obtains 5 further solutions of (2.25) with (2.27) for $S \subseteq N$ with $|S| = 2$ which are given by $u = (\frac{1}{3}, 0, 0, \frac{1}{3}, 0)^T$, $(\frac{1}{3}, 0, 0, 0, \frac{1}{3})^T$, $(0, \frac{1}{3}, \frac{1}{3}, 0, 0)^T$, $(0, \frac{1}{3}, 0, \frac{1}{3}, 0)^T$ and $(0, \frac{1}{3}, 0, 0, \frac{1}{3})^T$.

2.1.7 Zero-Sum Evolution Matrix Games

In this section we consider evolution matrix games with an antisymmetric payoff matrix $A = (a_{jk})$, i.e., $A = -A^T$ which is equivalent to

$$a_{ij} = -a_{ji}, \quad \text{for } i, j = 1, \dots, n \text{ with } i \neq j,$$
$$a_{ii} = 0 \text{ for } i = 1, \dots, n.$$

Such a game is called a zero-sum game.

Lemma 4. *If A is an antisymmetric matrix, then it follows that*

$$uAu^T = 0 \text{ for every } u \in \Delta.$$

Proof.

$$\begin{aligned} A &= -A^T \text{ implies} \\ uAu^T &= (uAu^T)^T = uA^T u^T = -(uAu^T), \\ \text{hence } uAu^T &= 0. \end{aligned}$$

□

Lemma 5. *Let A be antisymmetric. If $u^* \in \Delta$ is an evolutionarily stable Nash equilibrium, then $u^* \in \{e_1, e_2, \ldots, e_n\}$.*

Proof. Let us assume that $u^* \notin \{e_1, e_2, \ldots, e_n\}$. Then it follows that $|S(u^*)| > 1$, i.e., there exists a $k \in S(u^*)$ such that $e_k \neq u^*$. Lemma 1 implies that

$$e_k A u^{*T} = u^* A u^{*T} = 0.$$

Since u^* is evolutionarily stable, it follows that

$$u^* A e_k^T > e_k A e_k^T = 0.$$

On the other hand we have

$$u^* A e_k^T = (u^* A e_k^T)^T = e_k A^T u^{*T} = -e_k A u^{*T} = 0$$

which is a contradiction. Hence the assumption is false. □

Theorem 2.7 *Let A be antisymmetric. If e_l is an evolutionarily stable Nash equilibrium, then e_l is the only one.*

Proof. Let e_k, $k \neq l$, be another evolutionarily stable Nash equilibrium. Then we have

$$e_k A e_l^T \leq e_l A e_l^T = 0 \tag{2.30}$$

and

$$e_l A e_k^T \leq e_k A e_k^T = 0. \tag{2.31}$$

From (2.31) we infer

$$e_l A e_k^T = (e_l A e_k^T)^T = e_k A^T e_l^T = -e_k A e_l^T \leq 0$$

which implies

$$e_k A e_l^T \geq 0.$$

From (2.30) it therefore follows that

$$e_k A e_l^T = 0.$$

Since $e_l \in$ ESS, this implies

$$e_l A e_k^T > e_k A e_k^T = 0$$

which contradicts (2.31). □

Theorem 2.8 *Let A be antisymmetric. Then e_k is an evolutionarily stable Nash equilibrium, if and only if*

$$a_{ik} < 0 \textit{ for all } i = 1, \ldots, n \textit{ with } i \neq k. \tag{2.32}$$

Proof. 1. Let e_k be an evolutionarily stable Nash equilibrium. Then it follows

$$a_{ik} \leq a_{kk} = 0 \text{ for all } i = 1, \ldots, n$$

and, if

$$a_{ik} = 0 \text{ for some } i \neq k, \text{ then } a_{ik} > a_{ii} = 0.$$

Since $a_{ki} > 0$ implies $a_{ik} < 0$, $a_{ik} = 0$ for some $i \neq k$ cannot occur. This shows that (2.32) must hold.

2. Conversely, let (2.32) be true. Then, for every $u \in \Delta$ with $u \neq e_k$, it follows that

$$u A e_k^T = \sum_{i=1}^{n} u_i a_{ik} = \sum_{i \neq k} u_i a_{ik} < 0 = e_k A e_k^T$$

which implies that e_k is an evolutionarily stable Nash equilibrium.

□

In order to find out whether there exists an evolutionarily stable Nash equilibrium one has to check whether there exists a $k \in \{1, \ldots, n\}$ such that (2.32) is satisfied. According to Theorem 2.7 there can only be one such k which can be easily verified.

Because of

$$uAu^T = 0 \text{ for all } u \in \Delta$$

the dynamics introduced in Section 2.1.5 has to be modified for zero-sum games. In [4] it has been proposed to base it on the following recursion:

$$u_i^{k+1} = \frac{e_i A u^{k^T} + C}{C} u_i^k, \; i = 1, \dots, n, \tag{2.33}$$

where $C > 0$ is a constant with

$$C + e_i A u^T \geq 0 \text{ for all } e_i \text{ and } u \in \Delta,$$

which is interpreted as background fitness.

If we define

$$f_A(u)_i = \frac{e_i A u^T + C}{C} u_i \text{ for } i = 1, \dots, n \text{ and } u \in \Delta, \tag{2.34}$$

then $f_A : \Delta \to \Delta$ and the recursion (2.33) can be written in the form

$$u^{k+1} = f_A(u^k).$$

Further, every population state e_i is a fixed point of f_A.

Now we can prove the following

Theorem 2.9 *Let A be antisymmetric. If $u^* \in \Delta$ is an evolutionarily stable Nash equilibrium, then u^* is an asymptotically stable fixed point of f_A defined by* (2.34).

Proof. By Lemma 5 we infer that $u^* \in \{e_1, e_2, \dots, e_n\}$. Hence there exists $k \in \{1, \dots, n\}$ such that $u^* = e_k$ (by Theorem 2.7 e_k is the only one) and from Theorem 2.8 it follows that

$$a_{ik} < 0 \text{ for all } i = 1, \dots, n \text{ with } i \neq k.$$

This implies

$$e_k A u^T = \sum_{i=1}^{n} a_{ki} u_i = -\sum_{i=1}^{n} a_{ik} u_i > 0 \text{ for all } u \in \Delta \text{ with } u \neq e_k$$

and in turn

$$f_A(u)_k > u_k \text{ for all } u \in \Delta \text{ with } 0 < u_k < 1.$$

If we put $G = \{u \in \Delta | u_k > 0\}$ and define a Lyapunov function $V : \mathbb{R}^n \to \mathbb{R}$ by $V(u) = 1 - u_k$, then it follows that, for all $u \in G$,

$$V(u) \geq 0 \text{ and } V(u) = 0 \iff u = e_k$$

and, for all $u \in G$ with $u \neq e_k$,

$$V(f_A(u)) - V(u) = 1 - f_A(u)_k - (1 - u_k) = u_k - f_A(u)_k < 0.$$

This implies by Satz 5.8 in [3] (see also Section A.3) that $u^* = e_k$ is an asymptotically stable fixed point of f_A. □

In addition it follows from $f_A(G) \subseteq G$ that

$$\lim_{l \to \infty} f_A^l(u) = e_k \text{ for all } u \in G.$$

Conversely we can prove

Theorem 2.10 *Let A be antisymmetric and let e_k be an attractive fixed point of f_A. Then e_k is an evolutionarily stable Nash equilibrium.*

Proof. Since e_k is an attractor with respect to f_A, there exists a relatively open neighbourhood $U \subseteq \Delta$ of e_k with

$$\lim_{t \to \infty} f_A^t(u) = e_k \text{ for all } u \in U. \tag{2.35}$$

Now let $j \in \{1, \ldots, n\} \setminus \{k\}$. Then we choose $u = (u_1, \ldots, u_n) \in U$ with $u_j > 0$, $u_k > 0$, and $u_i = 0$ for $i = 1, \ldots, n$ with $i \neq j$ and $i \neq k$.

Then we obtain for every $t \in \mathbb{N}_0$

$$\begin{aligned} f_A^t(u)_i &= 0 \text{ for } i = 1, \ldots, n \text{ with } i \neq j \text{ and } i \neq k, \\ f_A^{t+1}(u)_k &= \frac{a_{kj} f_A^t(u)_j + C}{C} f_A^t(u)_k, \\ f_A^{t+1}(u)_j &= \frac{a_{jk} f_A^t(u)_k + C}{C} f_A(u)_j. \end{aligned}$$

This implies

$$f_A^{t+1}(u)_j - f_A^t(u)_j = \frac{f_A^t(u)_j f_A^t(u)_k}{C} a_{jk}.$$

Let us assume $a_{jk} \geq 0$.

Then it follows that

$$f_A^{t+1}(u)_j \geq f_A^t(u)_j \text{ for all } t \in \mathbb{N}_0$$

which implies

$$\lim_{t\to\infty} f_A^t(u)_j \geq u_j \geq 0.$$

This, however, contradicts (2.35) which shows that $a_{jk} < 0$. Since this holds true for every $j \in \{1, \ldots, n\} \setminus \{k\}$, it follows from Theorem 2.8 that e_k is an evolutionarily stable Nash equilibrium. □

2.2 Evolution-Bi-Matrix-Games for two Populations

2.2.1 The Game and Evolutionarily Stable Equilibria

We consider two populations that fight against each other in the struggle of life. We assume that the first population applies the strategies $I_1, I_2, \ldots, I_m$ and the second the strategies $J_1, J_2, \ldots, J_n$. If an individual of the first population that applies I_i meets an individual of the second that applies J_j the I_i-individual is given a payoff $a_{ij} \in \mathbb{R}$ by the J_j-individual and the J_j-individual is given a payoff $b_{ji} \in \mathbb{R}$ by the I_i-individual. All the payoffs then can be represented by the two matrices

$$A = (a_{ij})_{\substack{i=1,\ldots,m\\ j=1,\ldots,n}} \text{ and } B = (b_{ji})_{\substack{i=1,\ldots,m\\ j=1,\ldots,n}}.$$

These define a so called bi-matrix-game whose strategy sets are given by

$$S_1 = \left\{ u = (u_1, \ldots, u_m) | 0 \leq u_i \leq 1,\ i = 1, \ldots, m,\ \sum_{i=1}^{m} u_i = 1 \right\}$$

and

$$S_2 = \left\{ v = (v_1, \ldots, v_n) | 0 \leq v_j \leq 1,\ j = 1, \ldots, n,\ \sum_{j=1}^{n} v_j = 1 \right\}.$$

If two strategies $u \in S_1$ and $v \in S_2$ are given, then the average payoff of v to u is defined by

$$uAv^T = \sum_{i=1}^{m} \sum_{j=1}^{n} a_{ij} u_i v_j$$

and the average payoff of u to v by

$$vBu^T = \sum_{j=1}^{n} \sum_{i=1}^{m} b_{ji} u_i v_j.$$

Definition. *A pair* $(\hat{u}, \hat{v}) \in S_1 \times S_2$ *of strategies is called a Nash equilibrium, if*

$$\hat{u}A\hat{v}^T \geq uA\hat{v}^T \textit{ for all } u \in S_1$$
$$\textit{and} \tag{2.36}$$
$$\hat{v}B\hat{u}^T \geq vB\hat{u}^T \textit{ for all } v \in S_2.$$

In words this means that a deviation from $\hat{u}$ or $\hat{v}$ does not lead to a higher average payoff.

Let us demonstrate this by an example which is taken from [2]. We consider the "fight of sexes" in a population of animals. So the two populations we consider are the males and the females. The males have the two strategies of being faithful (I_1) and of being unfaithful (I_2) and the females have the two strategies of being willing (J_1) and of being resistant (J_2).

Let us assume that for every successfully grown up child the parents get both 15 points. The costs for growing up a child are assumed to be -20 points and the costs for a long time of "engagement" are assumed to be -3 points.

If a faithful male meets a resistant female, then the payoff for both is 2 points, namely 15 (for the child) -10 (for the shared costs of growing up) -3 (for the long time of "engagement"). If a faithful male meets a willing female, then they save the long time of "engagement" and get both 5 points. If an unfaithful male meets a willing female, then he gets 15 points and she $15 - 20 = -5$ points. If an unfaithful male meets a resistant female, both get 0 points. This leads to the two payoff matrices

$$A = \begin{pmatrix} 5 & 2 \\ 15 & 0 \end{pmatrix} \text{ and } B = \begin{pmatrix} 5 & -5 \\ 2 & 0 \end{pmatrix}$$

for the males and females, respectively.

The strategy sets are given by

$$S_1 = \{u = (u_1, u_2) | 0 \leq u_1, u_2 \leq 1 \text{ and } u_1 + u_2 = 1\}$$

and

$$S_2 = \{v = (v_1, v_2) | 0 \leq v_1, v_2 \leq 1 \text{ and } v_1 + v_2 = 1\}.$$

The first condition in (2.36) is equivalent with

$$\hat{u}A\hat{v}^T - uA\hat{v}^T = (2 - 12\hat{v}_1)(\hat{u}_1 - u_1) \geq 0 \text{ for all } u \in S_1$$

which is equivalent with

$$\hat{v}_1 = \frac{1}{6},\ \hat{v}_2 = \frac{5}{6} \text{ which implies } \hat{u}A\hat{v}^T = uA\hat{v}^T \text{ for all } u \in S_1$$

and the second condition in (2.36) is equivalent with

$$\hat{v}B\hat{u}^T - vB\hat{u}^T = (8\hat{u}_1 - 5)(\hat{v}_1 - v_1) \geq 0 \text{ for all } v \in S_2$$

which is equivalent with

$$\hat{u}_1 = \frac{5}{8},\ \hat{u}_2 = \frac{3}{8} \text{ which implies } \hat{v}B\hat{u}^T = vB\hat{u}^T \text{ for all } v \in S_2.$$

The only Nash equilibrium is therefore given by $(\hat{u}, \hat{v}) \in S_1 \times S_2$ with $\hat{u} = \left(\frac{5}{8}, \frac{3}{8}\right)$ and $\hat{v} = \left(\frac{1}{6}, \frac{5}{6}\right)$ and the corresponding average payoffs are

$$\hat{u}A\hat{v}^T = \frac{5}{2} \text{ and } \hat{v}B\hat{u}^T = \frac{5}{4}.$$

In this example we have the special case of a Nash equilibrium $(\hat{u}, \hat{v}) \in S_1 \times S_2$ with

$$\hat{u}_i > 0 \text{ for all } i = 1, \ldots, m \text{ and } \hat{v}_j > 0 \text{ for all } j = 1, \ldots, n. \tag{2.37}$$

In such a case it follows

$$e_i A\hat{v}^T = \hat{u}A\hat{v}^T \text{ for all } i = 1, \ldots, m \left(\text{where } e_i = (0, \ldots, \overset{i}{1}, \ldots, 0) \right)$$

and

$$e_j B\hat{u}^T = \hat{v}B\hat{v}^T \text{ for all } j = 1, \ldots, n \left(\text{where } e_j = (0, \ldots, \overset{j}{1}, \ldots, 0) \right). \tag{2.38}$$

This is a consequence of the following

Theorem 2.11 *If the pair $(\hat{u}, \hat{v}) \in S_1 \times S_2$ is a Nash equilibrium, then it follows that*

$$\begin{gathered} e_i A\hat{v}^T = \hat{u}A\hat{v}^T \textit{ for all } i = 1, \ldots, m \textit{ with } \hat{u}_i > 0 \\ \textit{and} \\ e_j B\hat{u}^T = \hat{v}B\hat{v}^T \textit{ for all } j = 1, \ldots, n \textit{ with } \hat{v}_j > 0. \end{gathered} \tag{2.39}$$

Proof. The first inequality of (2.36) is equivalent to

$$\hat{u}A\hat{v}^T = \max_{i=1,\dots,m} e_i A\hat{v}^T . \tag{2.40}$$

If we define

$$e_{i_0}A\hat{v}^T = \max_{i=1,\dots,m} e_i A\hat{v}^T ,$$

then (2.40) implies

$$\sum_{\hat{u}_i>0} \hat{u}_i \left(e_i A\hat{v}^T - e_{i_0}A\hat{v}^T\right) = 0$$

which is equivalent with the first equation in (2.39). The proof of the necessity of the second equation in (2.39) for $(\hat{u}, \hat{v})$ to be a Nash equilibrium is the same. □

In the above example we have

$$A\hat{v}^T = \begin{pmatrix} 5\hat{v}_1 + 2\hat{v}_2 \\ 15\hat{v}_1 \end{pmatrix} = \begin{pmatrix} \frac{5}{2} \\ \frac{5}{2} \end{pmatrix} \text{ and } B\hat{u}^T = \begin{pmatrix} 5\hat{u}_1 - 5\hat{u}_2 \\ 2\hat{u}_1 \end{pmatrix} = \begin{pmatrix} \frac{5}{4} \\ \frac{5}{4} \end{pmatrix}.$$

Under rational behavior (2.36) would be a condition that ensures stability in the sense that none of the two populations deviates from its strategy $\hat{u}$ and $\hat{v}$, respectively. However, without rational behavior stability has to be guaranteed by an additional condition. In analogy to the concept of evolutionarily stability in the case of one population we make the following

Definition. *A Nash equilibrium* $(\hat{u}, \hat{v}) \in S_1 \times S_2$ *is called evolutionarily stable, if*

$$\hat{u}A\hat{v}^T + \hat{v}B\hat{u}^T = uA\hat{v}^T + vB\hat{u}^T \textit{ for some } (u, v) \in S_1 \times S_2 \textit{ with } (u, v) \neq (\hat{u}, \hat{v})$$

implies

$$\hat{u}Av^T + \hat{v}Bu^T > uAv^T + vBu^T .$$

The following theorem shows that evolutionarily stable Nash equilibria can only be pairs of pure strategies.

Theorem 2.12 *If* $(\hat{u}, \hat{v}) \in S_1 \times S_2$ *is an evolutionarily stable Nash equilibrium, then*

$$\hat{u} \in \{e_1, \dots, e_m\} \textit{ and } \hat{v} \in \{e_1, \dots, e_n\} .$$

Proof. If $(\hat{u}, \hat{v}) \in S_1 \times S_2$ is a Nash equilibrium, it follows from Theorem 2.11 that

$$\hat{u}A\hat{v}^T + \hat{v}B\hat{u}^T = uA\hat{v}^T + vB\hat{u}^T$$

for all $u \in S_1$ with $S(u) \subseteq S(\hat{u})$ and all $v \in S_2$ with $S(v) \subseteq S(\hat{v})$.
Let us assume that $\hat{u} \notin \{e_1, \ldots, e_m\}$.
Then we choose $u \in S$, with $S(u) \subsetneq S(\hat{u})$ and $v = \hat{v}$ and conclude $(u, \hat{v}) \neq (\hat{u}, \hat{v})$ as well as

$$\hat{u}A\hat{v}^T + \hat{v}B\hat{u}^T = uA\hat{v}^T + \hat{v}B\hat{u}^T.$$

However, if $(\hat{u}, \hat{v})$ is evolutionarily stable, then the last equation implies

$$\hat{u}A\hat{v}^T + \hat{v}Bu^T > uA\hat{v}^T + \hat{v}Bu^T$$

which leads to a contradiction. Hence $\hat{u} \in \{e_1, \ldots, e_m\}$. □

In a similar way one shows that $\hat{v} \in \{e_1, \ldots, e_n\}$.

The Nash equilibrium in the above example can therefore not be evolutionarily stable.

2.2.2 A Dynamical Treatment of the Game

In the following we assume that

$$uAv^T > 0 \text{ and } vBu^T > 0 \text{ for all } u \in S_1 \text{ and } v \in S_2.$$

Under this assumption we define a mapping $f = (f_1, f_2) : S_1 \times S_2 \to S_1 \times S_2$ by

$$f_1(u, v)_i = \frac{e_i A v^T}{uAv^T} u_i \text{ for } i = 1, \ldots, m$$

and

$$f_2(u, v)_j = \frac{e_j B u^T}{vBu^T} v_j \text{ for } j = 1, \ldots, n.$$

This mapping is continuous and by Brouwer's fixed point theorem it has a fixed point. Now let $(\hat{u}, \hat{v}) \in S_1 \times S_2$ be a fixed point of f. Then it follows that

$$\begin{aligned} &e_i A\hat{v}^T = \hat{u}A\hat{v}^T \text{ for all } i \in S(\hat{u}) = \{i | \hat{u}_i > 0\} \\ &\text{and} \\ &e_j B\hat{u}^T = \hat{v}B\hat{u}^T \text{ for all } j \in S(\hat{v}) = \{j | \hat{v}_j > 0\}. \end{aligned} \tag{2.41}$$

Conversely, if these two conditions are satisfied, then $(\hat{u}, \hat{v}) \in S_1 \times S_2$ is a fixed point of f.

Since by Theorem 2.11 these two conditions are necessary for $(\hat{u}, \hat{v})$ to be a Nash equilibrium, it follows that $(\hat{u}, \hat{v})$ is a fixed point of f, if $(\hat{u}, \hat{v})$ is a Nash equilibrium.

Conversely we have the

Theorem 2.13 *If $(\hat{u}, \hat{v}) \in S_1 \times S_2$ is a fixed point of f with*

$$S(\hat{u}) = \{1, \ldots, m\} \text{ and } S(\hat{v}) = \{1, \ldots, n\}, \tag{2.42}$$

then $(\hat{u}, \hat{v})$ is a Nash equilibrium.

Proof. If $(\hat{u}, \hat{v}) \in S_1 \times S_2$ is a fixed point of f with (2.42), then (2.41) implies that (2.38) must be satisfied and in turn

$$uA\hat{v}^T = \hat{u}A\hat{v}^T \text{ for all } u \in S_1$$

and

$$vB\hat{u}^T = \hat{v}B\hat{u}^T \text{ for all } v \in S_2$$

which shows that $(\hat{u}, \hat{v})$ is a Nash equilibrium. □

Next we prove the

Theorem 2.14 *If $(\hat{u}, \hat{v}) \in S_1 \times S_2$ is an attractive fixed point of f, i.e. a fixed point which is an attractor, then $(\hat{u}, \hat{v})$ is a Nash equilibrium.*

Proof. If $(\hat{u}, \hat{v}) \in S_1 \times S_2$ satisfies the condition (2.42), then the assertion follows from Theorem 2.13.

If $S(\hat{u}) \neq \{1, \ldots, m\}$ or $S(\hat{v}) \neq \{1, \ldots, n\}$, then (2.41) is satisfied. If we then show

$$e_i A\hat{v}^T \leq \hat{u}A\hat{v}^T \text{ for all } i \notin S(\hat{u})$$

and

$$e_j B\hat{u}^T \leq \hat{v}B\hat{u}^T \text{ for all } j \notin S(\hat{v}),$$

it follows that $(\hat{u}, \hat{v})$ is a Nash equilibrium.

Let us assume that there is some $k \in \{1, \ldots, m\} \setminus S(\hat{u})$ with

$$e_k A\hat{v}^T > \hat{u}A\hat{v}^T. \tag{2.43}$$

Since the function $g(u, v) = e_k A v^T - u A \hat{v}^T$, $(u, v) \in S_1 \times S_2$ is continuous, there exists some $\epsilon_1 > 0$ such that

$$\begin{aligned} & e_k A v^T > u A v^T \text{ for all } u \in S_1 \text{ and } v \in S_2 \\ & \text{with } \|u - \hat{u}\|_2 < \epsilon_1 \text{ and } \|v - \hat{v}\|_2 < \epsilon_1. \end{aligned} \tag{2.44}$$

Since $(\hat{u}, \hat{v})$ is an attractor, there exists some $\epsilon_2 > 0$ such that

$$\begin{aligned} & \lim_{t\to\infty} f^t(u, v) = (\hat{u}, \hat{v}) \text{ for all } u \in S_1 \text{ and } v \in S_2 \\ & \text{with } \|u - \hat{u}\|_2 < \epsilon_2 \text{ and } \|v - \hat{v}\|_2 < \epsilon_2. \end{aligned} \tag{2.45}$$

This implies that for every pair

$$(u, v) \in W = \{(u, v) \in S_1 \times S_2 | \; \|u - \hat{u}\|_2 < \epsilon \text{ and } \|v - \hat{v}\|_2 < \epsilon\}$$

with $\epsilon = \min(\epsilon_1, \epsilon_2)$ there exists a $T_\epsilon \in \mathbb{N}$ such that

$$\|f_1^t(u, v) - \hat{u}\|_2 < \epsilon \text{ for all } t \geq T_\epsilon.$$

From (2.44) we infer therefore

$$f_1^{t+1}(u, v)_k > f_1^t(u, v)_k > 0 \text{ for all } (u, v) \in W \text{ and all } t \geq T_\epsilon. \tag{2.46}$$

On the other hand it follows from (2.45) that

$$\lim_{t\to\infty} f_1^t(u, v)_k = \hat{u}_k = 0 \text{ for all } (u, v) \in W$$

which contradicts (2.46). Therefore the assumption (2.43) is false and the Theorem 2.14 is proved. □

In the following we concentrate on Nash equilibria which are pairs of pure strategies, since in view of Theorem 2.12 they are the only Nash equilibria which are evolutionarily stable.

For these we can prove the

Theorem 2.15 *Let $(e_k, e_l) \in S_1 \times S_2$ be a Nash equilibrium such that there exists a relatively open subset $U \subseteq S_1$ with $e_k \in U$ and a relatively open subset $V \subseteq S_2$ with $e_l \in V$ such that*

$$\begin{aligned} & e_k A v^T > u A v^T \textit{ and } e_l B u^T > v B u^T \\ & \textit{for all } (u, v) \in U \times V \textit{ with } u \neq e_k \textit{ and } v \neq e_l. \end{aligned} \tag{2.47}$$

Then (e_k, e_l) is an asymptotically stable fixed point of f.

Proof. From (2.47) it follows that

$$\begin{gathered} f_1(u,v)_k > u_k \text{ and } f_2(u,v)_l > v_l \\ \text{for all } (u,v) \in U \times V \text{ with } u \neq e_k \text{ and } v \neq e_l. \end{gathered} \tag{2.48}$$

If we now define a continuous function $V : S_1 \times S_2 \to \mathbb{R}$ by

$$V(u,v) = 2 - u_k - v_l \text{ for } (u,v) \in S_1 \times S_2,$$

then it follows that

$$\begin{gathered} V(f(u,v)) - V(u,v) = 2 - f_1(u,v)_k - f_2(u,v)_l - 2 + u_k + v_l < 0 \\ \text{for all } (u,v) \in U \times V \text{ with } u \neq e_k \text{ and } v \neq e_l. \end{gathered}$$

and

$$V(e_k, e_l) = 0 \text{ and } V(u,v) > 0, \quad \text{if } (u,v) \neq (e_k, e_l).$$

By Satz 5.8 in [3] (see also Section A.3) this implies that (e_k, e_l) is an asymptotically stable fixed point of f. □

In addition we infer from the proof of Satz 5.8 in [3] the existence of a relatively open subset $W \subseteq U \times V$ with $(e_k, e_l) \in W$ such that $f(W) \subseteq W$ and

$$(e_k, e_l) = \lim_{t\to\infty} f^t(u,v) \text{ for all } (u,v) \in W. \tag{2.49}$$

Now let us choose arbitrarily some pair $(e^\circ, v^\circ) \in W$ and define a sequence $((u^t, v^t))_{t\in\mathbb{N}_0}$ in W via

$$(u^{t+1}, v^{t+1}) = (f_1(u^t, v^t), f_2(u^t, v^t)), \ t \in N_0.$$

Then it follows from (2.48) that

$$\left.\begin{aligned} &f_1^{t+1}(u^\circ, v^\circ)_k \geq f_1^t(u^\circ, v^\circ)_k \\ \text{and} \quad & \\ &f_2^{t+1}(u^\circ, v^\circ)_l \geq f_2^t(u^\circ, v^\circ)_l \end{aligned}\right. \quad \text{for all } t \in \mathbb{N}_0$$

which implies

$$f_1^t(u^\circ, v^\circ)_k \nearrow 1 \text{ and } f_2^t(u^\circ, v^\circ)_l \nearrow 1 \text{ for } t \to \infty,$$

hence,

$$\sum_{i\neq k} f_1^t(u^\circ, v^\circ)_i \searrow 0 \text{ and } \sum_{j\neq l} f_2^t(u^\circ, v^\circ)_j \searrow 0 \text{ for } t \to \infty.$$

As a result we obtain

$$\lim_{t\to\infty} f^t(u^\circ, v^\circ) = (e_k, e_l).$$

An example: Let $m = 2,\ n = 3,$

$$A = \begin{pmatrix} 5 & 3 & 1 \\ 6 & 4 & 2 \end{pmatrix} \text{ and } B = \begin{pmatrix} 4 & 3 \\ 5 & 2 \\ 6 & 1 \end{pmatrix}.$$

Then $(e_2, e_1) \in S_1 \times S_2$ is a Nash equilibrium where

$$S_1 = \left\{ u \in \mathbb{R}^2 | u_1 \geq 0,\ u_2 \geq 0,\ u_1 + u_2 = 1 \right\}$$

and

$$S_2 = \left\{ v \in \mathbb{R}^3 | v_1 \geq 0,\ v_2 \geq 0,\ v_3 \geq 0,\ v_1 + v_2 + v_3 = 1 \right\}.$$

Further we obtain

$$e_2 A v^T = (0, 1) \begin{pmatrix} 5v_1 + 3v_2 + v_3 \\ 6v_1 + 4v_2 + 2v_3 \end{pmatrix} = 6v_1 + 4v_2 + 2v_3$$

and

$$\begin{aligned} uAv^T &= (u_1, u_2) \begin{pmatrix} 5v_1 + 3v_2 + v_3 \\ 6v_1 + 4v_2 + 2v_3 \end{pmatrix} \\ &= u_1(5v_1 + 3v_2 + v_3) + u_2(6v_1 + 4v_2 + 2v_3) = 6v_1 + 4v_2 + 2v_3 - u_1, \end{aligned}$$

hence,

$$e_2 A v^T - u A v^T = u_1 > 0, \quad \text{if } u_1 > 0.$$

Finally we obtain

$$e_1 B u^T = (1, 0, 0) \begin{pmatrix} 4u_1 + 3u_2 \\ 5u_1 + 2u_2 \\ 6u_1 + u_2 \end{pmatrix} = 4u_1 + 3u_2 = 3 + u_1$$

and

$$\begin{aligned} vBu^T &= (v_1, v_2, v_3) \begin{pmatrix} 4u_1 + 3u_2 \\ 5u_1 + 2u_2 \\ 6u_1 + u_2 \end{pmatrix} \\ &= v_1(4u_1 + 3u_2) + v_2(5u_1 + 2u_2) + v_3(6u_1 + u_2) \\ &= 3v_1 + 2v_2 + v_3 + u_1(v_1 + 3v_2 + 5v_3), \end{aligned}$$

hence,

$$\begin{aligned} e_1 B u^T - v B u^T &= 3 + u_1 - 3v_1 - 2v_2 - v_3 - u_1(v_1 + 3v_2 + 5v_3) \\ &= 3 + u_1(-2v_2 - 4v_4) - 3v_1 - 2v_2 - v_3 \\ &> 3 - 3v_1 - 3v_2 - 3v_3 = 0, \quad \text{if } u_1 < \tfrac{1}{2}. \end{aligned}$$

This implies that the conditions (2.47) are satisfied for

$$U = \left\{ u \in S_1 | 0 \le u_1 < \frac{1}{2} \right\} \text{ and } V = S_2.$$

By Theorem 2.15 it follows that (e_2, e_1) is an asymptotically stable fixed point of f. Further (2.49) holds true for $k = 2,\ l = 1$ and $W = U \times V$.

2.2.3 Existence and Iterative Calculation of Nash Equilibria

For every pair $(u, v) \in S_1 \times S_2$ we define

$$\varphi_{1j}(u, v) = \max(0, e_j A v^T - u A v^T) \text{ for } j = 1, \ldots, m$$

and

$$\varphi_{2k}(u, v) = \max(0, e_k B u^T - v B u^T) \text{ for } k = 1, \ldots, n$$

and with these functions we define a mapping $f = (f_1, f_2) : S_1 \times S_2 \to S_1 \times S_2$ by

$$f_1(u, v)_j = \frac{1}{1 + \sum\limits_{i=1}^{m} \varphi_{1i}(u, v)} (u_j + \varphi_{1j}(u, v)) \text{ for } j = 1, \ldots, m$$

and

$$f_2(u, v)_k = \frac{1}{1 + \sum\limits_{l=1}^{n} \varphi_{2l}(u, v)} (v_k + \varphi_{2k}(u, v)) \text{ for } k = 1, \ldots, n.$$

This mapping is continuous and by Brouwer's fixed point theorem it has a fixed point. For every fixed point of f we have the following

Theorem 2.16 $(\hat{u}, \hat{v}) \in S_1 \times S_2$ *is a fixed point of* f, *if and only if* $(\hat{u}, \hat{v})$ *is a Nash equilibrium.*

Proof. 1. Let $(\hat{u}, \hat{v}) \in S_1 \times S_2$ be a fixed point of f, i.e.,

$$f_1(\hat{u}, \hat{v})_j = \hat{u}_j \text{ for } j = 1, \dots, m \text{ and } f_2(\hat{u}, \hat{v})_k = \hat{v}_k \text{ for } k = 1, \dots, n.$$

Let $\hat{u}_j > 0$ for some $j \in \{1, \dots, m\}$ (at least one must exists).

Then we choose $j_1 \in \{1, \dots, m\}$ such that

$$e_{j_1} A\hat{v}^T = \min\{e_j A\hat{v}^T | \hat{u}_j > 0\}$$

and conclude $\varphi_{1j_1}(\hat{u}, \hat{v}) = 0$ because of $e_{j_1} A\hat{v}^T \leq \hat{u}A\hat{v}^T = \sum_{\hat{u}_j > 0} \hat{u}_j e_j A\hat{v}^T$.

This in turn implies

$$\varphi_{1i}(\hat{u}, \hat{v}) = 0 \text{ for } i = 1, \dots, m.$$

Similarly one proves that

$$\varphi_{2l}(\hat{u}, \hat{v}) = 0 \text{ for } l = 1, \dots, n.$$

The last two conditions are equivalent to $(\hat{u}, \hat{v})$ being a Nash equilibrium.

2. Conversely, let $(\hat{u}, \hat{v}) \in S_1 \times S_2$ be a Nash equilibrium. Then the last two conditions hold true which implies that $(\hat{u}, \hat{v})$ is a fixed point of f.

□

According to Theorem 2.16 the existence of Nash equilibria is ensured. But it also gives rise to an iterative method for calculating Nash equilibria. Let us demonstrate this by the following example (see Section 2.2.2):
Let $m = 2,\ n = 3$,

$$A = \begin{pmatrix} 5 & 3 & 1 \\ 6 & 4 & 2 \end{pmatrix} \text{ and } B = \begin{pmatrix} 4 & 3 \\ 5 & 2 \\ 6 & 1 \end{pmatrix}.$$

Then we obtain, for every $(u, v) \in S_1 \times S_2$,

$$\begin{aligned} e_1 Av^T &= 5v_1 + 3v_2 + v_3, \\ e_2 Av^T &= 6v_1 + 4v_2 + 2v_3, \\ uAv^T &= 6v_1 + 4v_2 + 2v_3 - u_1 \end{aligned}$$

which implies

$$e_1 Av^T - uAv^T = -1 + u_1 \leq 0$$

and

$$e_2 Av^T - uAv^T = u_1.$$

Therefore we have

$$\varphi_{11}(u,v) = 0 \text{ and } \varphi_{12}(u,v) = u_1 > 0$$
$$\text{for all } (u,v) \in S_1 \times S_2 \text{ with } 0 < u_1 \le 1$$

and further

$$f_1(u,v)_1 = \frac{u_1}{1+u_1},\ f_1(u,v)_2 = \frac{1}{1+u_1} = 1 - f_1(u,v)_1. \tag{2.50}$$

Next we obtain, for every $(u,v) \in S_1 \times S_2$,

$$\begin{aligned} e_1 B u^T &= 4u_1 + 3u_2 = 3 + u_1, \\ e_2 B u^T &= 5u_1 + 2u_2 = 2 + 3u_1, \\ e_3 B u^T &= 6u_1 + u_2 = 1 + 5u_1, \\ v B u^T &= 3v_1 + 2v_2 + v_3 + u_1(v_1 + 3v_2 + 5v_3) \end{aligned}$$

which implies

$$\begin{aligned} e_1 B u^T - v B u^T &= 3 - u_1(2v_2 + 4v_3) - 3v_1 - 2v_2 - v_3 \\ &= 3 - 3v_1 - 2(u_1+1)v_2 - (4u_1+1)v_3 > 0, \text{ if } u_1 < \frac{1}{2}, \\ e_2 B u^T - v B u^T &= 2 + u_1(2v_1 - 2v_3) - 3v_1 - 2v_2 - v_3 \\ &< 2 + v_1 - v_3 - 3v_1 - 2v_2 - v_3 = 0, \text{ if } v_1 > v_3 \text{ and } u_1 < \frac{1}{2}, \\ e_3 B u^T - v B u^T &= 1 + u_1(4v_1 + 2v_2) - 3v_1 - 2v_2 - v_3 \\ &< 1 + 2v_1 + v_2 - 3v_1 - 2v_2 - v_3 = 0, \text{ if } u_1 < \frac{1}{2}, \end{aligned}$$

hence,

$$\varphi_{21}(u,v) > 0,\ \varphi_{22}(u,v) = \varphi_{23}(u,v) = 0$$
$$\text{for all } (u,v) \in S_1 \times S_2 \text{ with } u_1 < \tfrac{1}{2} \text{ and } v_1 > v_3. \tag{2.51}$$

This implies

$$\begin{aligned} f_2(u,v)_1 &= \frac{v_1 + \varphi_{21}(u,v)}{1 + \varphi_{21}(u,v)} = 1 - f_2(u,v)_2 - f_2(u,v)_3, \\ f_2(u,v)_2 &= \frac{v_2}{1 + \varphi_{21}(u,v)}, \\ f_2(u,v)_3 &= \frac{v_3}{1 + \varphi_{21}(u,v)} \end{aligned} \tag{2.52}$$

for all $(u,v) \in S_1 \times S_2$ with $u < \frac{1}{2}$ and $v_1 > v_3$.

From (2.50), (2.51) and (2.52) it follows that

$$f_1(u,v)_1 < u_1,\ f_1(u,v)_2 > u_2,$$
$$f_2(u,v)_2 < v_2, f_2(u,v)_3 < v_3 \Rightarrow f_2(u,v)_1 > v_1$$
$$\text{for all } (u,v) \in U = \{(u,v) \in S_1 \times S_2 | 0 < u_1 < \frac{1}{2}, v_1 > v_3.\}$$

Further it follows that $f(U) \subseteq U$.

If we now choose arbitrarily some $(u^\circ, v^\circ) \in U$ and define a sequence $((u^t, v^t))_{t\in\mathbb{N}_0}$ in U via

$$(u^{t+1}, v^{t+1}) = f(u^t, v^t),\ t \in \mathbb{N}_0, \tag{2.53}$$

then it follows that

$$u_1^t \to 0,\ u_2^t \to 1,\ v_1^t \to 1,\ v_2^t \to 0,\ v_3^t \to 0 \ \text{(Exercise)}$$

hence $(u^t, v^t) \to (e_2, e_1) = f(e_2, e_1)$. By Theorem 2.16 (e_2, e_1) is a Nash equilibrium.

This example is a special case of the following general situation:

Let $U \subseteq S_1 \times S_2$ be such that there exists some $j_0 \in \{1, \ldots, m\}$ and some $k_0 \in \{1, \ldots, n\}$ with

$$\left.\begin{array}{c} \varphi_{1j}(u,v) = 0 \text{ for all } j \neq j_0, \varphi_{1j_0} > 0 \\ \text{and} \\ \varphi_{2k}(u,v) = 0 \text{ for all } k \neq k_0, \varphi_{2k_0}(u,v) > 0 \end{array}\right\} \text{for all } (u,v) \in U.$$

Then it follows that

$$\begin{aligned}
f_1(u,v)_j &= \frac{u_j}{1 + \varphi_{1j_0}(u,v)} \text{ for } j \neq j_0, \\
f_1(u,v)_{j_0} &= \frac{u_{j_0} + \varphi_{1j_0}(u,v)}{1 + \varphi_{1j_0}(u,v)} = 1 - \sum_{j \neq j_0} f_1(u,v)_j, \\
f_2(u,v)_k &= \frac{v_k}{1 + \varphi_{2k_0}(u,v)} \text{ for } k \neq k_0, \\
f_2(u,v)_{k_0} &= \frac{v_{k_0} + \varphi_{2k_0}(u,v)}{1 + \varphi_{2k_0}(u,v)} = 1 - \sum_{k \neq k_0} f_2(u,v)_k
\end{aligned}$$

and in turn that

$$f_1(u,v)_j < u_j \text{ for all } j \neq j_0 \Rightarrow f_1(u,v)_{j_0} > u_{j_0},$$
$$f_2(u,v)_k < v_k \text{ for all } k \neq k_0 \Rightarrow f_2(u,v)_{k_0} > v_{k_0}$$

for all $(u, v) \in U$.

Assumption: $f(U) \subseteq U$.

If we now choose some $(u^0, v^0) \in U$ and define a sequence $((u^t, v^t))_{t\in\mathbb{N}_0}$ in U via (2.53) then it follows that

$$(u^t, v^t) \to (\hat{u}, \hat{v}) = f(\hat{u}, \hat{v}) \in \overline{U} \text{ and } (\hat{u}, \hat{v}) \text{ is an Nash equilibrium.}$$

This implies in particular that

$$\begin{aligned} e_{j_0}A\hat{v}^T - \hat{u}A\hat{v}^T &= \sum_{k=1}^{n} a_{j_0k}\hat{v}_k - \sum_{j=1}^{m} \hat{u}_j \sum_{k=1}^{n} a_{jk}\hat{v}_k \\ &= \sum_{j=1}^{m} \hat{u}_j \left(\sum_{k=1}^{n} a_{j_0k}\hat{v}_k - \sum_{k=1}^{n} a_{jk}\hat{v}_k \right) \\ &= \sum_{j\neq j_0} \hat{u}_j \sum_{k=1}^{n} (a_{j_0k} - a_{jk})\hat{v}_k \leq 0. \end{aligned}$$

Assumption 1:

$$a_{j_0k} - a_{jk} > 0 \text{ for all } j \neq j_0 \text{ and } k = 1, \ldots, n.$$

This implies

$$\hat{u}_j = 0 \text{ for all } j \neq j_0, \text{ hence } \hat{u}_{j_0} = 1, \text{ and therefore } \hat{u} = e_{j_0}.$$

This again implies

$$\begin{aligned} e_{k_0}Be_{j_0}^T - \hat{v}Be_{j_0}^T &= b_{k_0j_0} - \sum_{k\neq k_0} b_{kj_0}\hat{v}_k \\ &= \sum_{k\neq k_0} \left(b_{k_0j_0} - b_{kj_0} \right) \hat{v}_k \leq 0. \end{aligned}$$

Assumption 2:

$$b_{k_0j_0} - b_{kj_0} > 0 \text{ for all } k \neq k_0.$$

This implies $\hat{v}_k = 0$ for all $k \neq k_0$, hence $\hat{v}_{k_0} = 1$, and therefore $\hat{v} = e_{k_0}$. In the above example the Assumption 1 holds for $j_0 = 2$ and the Assumption 2 holds for $j_0 = 2$ and $k_0 = 1$.

2.2.4 A Direct Method for the Calculation of Nash Equilibria

We start with the following necessary and sufficient condition for a Nash equilibrium $(\hat{u}, \hat{v}) \in S_1 \times S_2$ which is equivalent to the definition (2.36), namely,

$$\hat{u}A\hat{v}^T = \max_{i=1,\dots,m} e_i A\hat{v}^T$$

and

$$\hat{v}B\hat{u}^T = \max_{j=1,\dots,n} e_j B\hat{u}^T \Leftrightarrow \hat{u}B^T\hat{v}^T = \max_{j=1,\dots,n} \hat{u}B^T e_j^T.$$

This condition again is equivalent to

$$\hat{u}_k > 0 \Rightarrow e_k A\hat{v}^T = \max_{i=1,\dots,m} e_i A\hat{v}^T = \hat{u}A\hat{v}^T$$

and

$$\hat{v}_l > 0 \Rightarrow \hat{u}B^T e_l^T = \max_{j=1,\dots,n} \hat{u}B^T e_j^T = \hat{v}B\hat{u}^T$$

and in turn to

$$(\hat{u}A\hat{v}^T)e^m \geq A\hat{v}^T,$$
$$\hat{u}\left(A\hat{v}^T - (\hat{u}A\hat{v}^T)e^m\right) = 0 \quad \text{with } e^m = (1, \dots, 1)^T \in \mathbb{R}^m$$

and

$$(\hat{v}B\hat{u}^T)e^n \geq B\hat{u}^T,$$
$$\hat{v}\left(B\hat{u}^T - (\hat{v}B\hat{u}^T)e^n\right) = 0 \quad \text{with } e^n = (1, \dots, 1)^T \in \mathbb{R}^n.$$

If one defines

$$\hat{\chi} = -A\hat{v}^T + (\hat{u}A\hat{v}^T)e^m \quad \text{and} \quad \hat{y} = -B\hat{u}^T + (\hat{v}B\hat{u}^T)e^n,$$

then it follows that

$$\begin{pmatrix} \hat{\chi} \\ \hat{y} \end{pmatrix} = \begin{pmatrix} 0 & -A \\ -B & 0 \end{pmatrix} \begin{pmatrix} \hat{u}^T \\ \hat{v}^T \end{pmatrix} + \begin{pmatrix} (\hat{u}A\hat{v}^T)e^m \\ (\hat{v}B\hat{u}^T)e^n \end{pmatrix} \geq \begin{pmatrix} \Theta_m \\ \Theta_n \end{pmatrix} \tag{2.54}$$

$$\text{and} \quad \hat{\chi}^T \hat{u}^T + \hat{y}^T \hat{v}^T = 0 \Leftrightarrow \hat{\chi}^T \hat{u}^T = 0 \quad \text{and} \quad \hat{y}^T \hat{v}^T = 0.$$

These two conditions are then necessary and sufficient for $(\hat{u}, \hat{v}) \in S_1 \times S_2$ to be a Nash equilibrium.
Now let $(u^T, v^T) \in \mathbb{R}^m_+ \times \mathbb{R}^n_+$, $u^T \neq \Theta_m$, $v^T \neq \Theta_n$ be a solution of

$$\begin{pmatrix} \chi \\ y \end{pmatrix} = \begin{pmatrix} 0 & -A \\ -B & 0 \end{pmatrix} \begin{pmatrix} u^T \\ v^T \end{pmatrix} + \begin{pmatrix} e^m \\ e^n \end{pmatrix} \geq \begin{pmatrix} \Theta_m \\ \Theta_n \end{pmatrix}, \tag{2.55}$$

$$\chi^T u^T = 0 \text{ and } y^T v^T = 0.$$

Then we have $(e^m)^T u^T > 0$ and $(e^n)^T v^T > 0$ and it follows that

$$(e^m)^T u^T = uAv^T \quad \text{and} \quad (e^n)^T v^T = vBu^T.$$

If we put

$$\hat{u} = \frac{1}{(e^m)^T u^T} u \quad \text{and} \quad \hat{v} = \frac{1}{(e^n)^T v^T} v, \tag{2.56}$$

then it follows that $(\hat{u}, \hat{v}) \in S_1 \times S_2$ and

$$-A\hat{v}^T + \frac{1}{(e^n)^T v^T} e^m = -A\hat{v}^T + (\hat{u}A\hat{v}^T)e^m \geq \Theta_m,$$

$$\hat{u} \underbrace{\left(-A\hat{v}^T + (\hat{u}A\hat{v}^T)e^m\right)}_{\hat{\chi}} = \frac{1}{(e^m)^T u^T} u \left(-A \frac{1}{(e^n)^T v^T} v^T + (\hat{u}A\hat{v}^T)e^m\right)$$

$$= \frac{1}{(e^m)^T u^T} \frac{1}{(e^n)^T v^T} u \underbrace{\left(-Av^T + e^m\right)}_{\chi} = 0.$$

Similarly it follows that

$$-B\hat{u}^T + (\hat{v}B\hat{u}^T)e^n \geq \Theta_n \quad \text{and} \quad \hat{v}\underbrace{\left(-B\hat{u}^T + (\hat{u}B\hat{v}^T)e^n\right)}_{y} = 0.$$

The conditions (2.54) are therefore satisfied and $(\hat{u}, \hat{v}) \in S_1 \times S_2$ is a Nash equilibrium.

So in order to find a solution $(\hat{u}, \hat{v}) \in S_1 \times S_2$ of (2.54) one has to find a solution $(u^T, v^T) \in \mathbb{R}^m_+ \times \mathbb{R}^n_+$, $u^T \neq \Theta_m$, $v^T \neq \Theta_n$ of (2.55) and to define $(\hat{u}, \hat{v})$ by (2.56).
For this purpose we represent the inequality systems in (2.55) by the following two tableaus:

		$-v_1$	$-v_2$	$\ldots$	$-v_n$
χ_1	1	a_{11}	a_{12}	$\ldots$	a_{1n}
χ_2	1	a_{21}	a_{22}	$\ldots$	a_{2n}
$\vdots$	$\vdots$	$\vdots$	$\vdots$	$\ddots$	$\vdots$
χ_m	1	a_{m1}	a_{m2}	$\ldots$	a_{mn}

		$-u_1$	$-u_2$	$\ldots$	$-u_m$
y_1	1	b_{11}	b_{21}	$\ldots$	b_{m1}
y_2	1	b_{12}	b_{22}	$\ldots$	b_{m2}
$\vdots$	$\vdots$	$\vdots$	$\vdots$	$\ddots$	$\vdots$
y_n	1	b_{1n}	b_{2n}	$\ldots$	b_{mn}

At the beginning of the solution procedure in the left tableau some v_{j_1} is exchanged with some χ_{i_1} (with the aid of a Jordan elimination step) such that in the new tableau the first column only exists of non-negative elements. Then in the right tableau the variable u_{i_1} is exchanged with some y_{j_2} such that the first colum of the new tableau only consists of non-negative elements. If $j_2 = j_1$, then the procedure stops and the positive components of the solution of (2.55) can be found in the first column of the corresponding tableau.

If $j_2 \neq j_1$, then in the left tableau the variable v_{j_2} is exchanged with some χ_{i_2} (or v_{i_2}) such that the first column of the new tableau only exists of non-negative elements. Then in the right tableau u_{i_2} is exchanged with some y_{j_3}

such that the first column of the new tableau only exists of non-negative elements. If $j_3 = j_1$, then the procedure stops and the positive components of the solution of (2.55) can be found in the first column of the corresponding tableau.
If $j_3 \neq j_1$ then the procedure is continued in the same way as above.

We demonstrate the procedure by the following example: Let $m = n = 3$ and

$$A = \begin{pmatrix} 2 & 2 & 0 \\ 0 & 3 & 0 \\ 3 & 0 & 1 \end{pmatrix} \quad \text{and} \quad B = \begin{pmatrix} 3 & 0 & 2 \\ 0 & 3 & 2 \\ 0 & 0 & 1 \end{pmatrix}.$$

Then the initial tableaus read:

		$-v_1$	$-v_2$	$-v_3$
x_1	1	2	2	0
x_2	1	0	3	0
x_3	1	$\boxed{3}$	0	1

		$-u_1$	$-u_2$	$-u_3$
y_1	1	3	0	0
y_2	1	0	3	0
y_3	1	2	2	$\boxed{1}$

If we exchange in the left tableau v_1 with x_3 and in the right tableau u_3 with y_3, then we obtain the tableaus

		$-x_3$	$-v_2$	$-v_3$
x_1	$\frac{1}{3}$	$-\frac{2}{3}$	2	$-\frac{2}{3}$
x_2	1	0	3	0
v_1	$\frac{1}{3}$	$\frac{1}{3}$	0	$\boxed{\frac{1}{3}}$

		$-u_1$	$-u_2$	$-y_3$
y_1	1	3	0	0
y_2	1	0	$\boxed{3}$	0
u_3	1	2	2	1

In this first step we have $j_1 = 1$, $i_1 = 3$ and $j_2 = 3 \neq j_1$. Therefore we exchange in the left tableau v_3 with v_1 and obtain the tableau

		$-\chi_3$	$-v_2$	$-v_1$
χ_1	1	0	2	2
χ_2	1	0	$\boxed{3}$	0
v_3	1	1	0	3

Here the procedure can stop and we obtain $u^T = v^T = (0, 0, 1)^T$ as solution of (2.55). One can also continue with an exchange of v_2 with χ_2 and u_2 with y_2. This leads to the two tableaus

		$-\chi_3$	$-\chi_2$	$-v_1$
χ_1	$\frac{1}{3}$	0	$-\frac{2}{3}$	2
v_2	$\frac{1}{3}$	0	$\frac{1}{3}$	0
v_3	1	1	0	3

		$-u_1$	$-y_2$	$-y_3$
y_1	1	3	0	0
u_2	$\frac{1}{3}$	0	$\frac{1}{3}$	0
u_3	$\frac{1}{3}$	2	$-\frac{2}{3}$	1

and delivers $u^T = (0, \frac{1}{3}, \frac{1}{3})^T$ and $v^T = (0, \frac{1}{3}, 1)^T$ as solution of (2.55).

The same solution is obtained, if one starts the procedure with $j_1 = 2$ or $j_1 = 3$.
Starting with $j_1 = 2$ one additionally obtains the solution $u^T = v^T = (0, \frac{1}{3}, 0)^T$.

The existence of a solution $(u^T, v^T) \in \mathbb{R}^m_+ \times \mathbb{R}^n_+, u^T \neq \Theta_n, v^T \neq \Theta_m$ of (2.55) necessarily required that

$$uAv^T = ((e^m)^T u^T) > 0 \text{ and } vBu^T = ((e^n)^T v^T) > 0.$$

This is guaranteed, if $A > 0_{m\times n}$ and $B > 0_{n\times m}$, i.e. all elements of A and B are positive.

This, however, can be assumed without loss of generality. In order to see that we choose some $k > 0$ such that

$$kE + A > 0_{m\times n} \text{ and } kE^T + B > 0_{n\times m}$$

where

$$E = \begin{pmatrix} 1 & \dots & 1 \\ \vdots & \ddots & \vdots \\ 1 & \dots & 1 \end{pmatrix} \in \mathbb{R}^{m\cdot n}.$$

Now let $(u^T, v^T) \in \mathbb{R}_+^m \times \mathbb{R}_+^n$, $u^T \neq \Theta_m$, $v^T \neq \Theta_n$ be a solution of

$$\begin{aligned} \chi &= -(kE + A)v^T + e^m \geq \Theta_m, \\ y &= -(kE^T + B)u^T + e^n \geq \Theta_n, \\ \chi^T u^T &= 0 \text{ and } y^T v^T = 0. \end{aligned} \tag{2.57}$$

Then it follows that

$$\begin{aligned} k &= \frac{1}{(e^n)^T v^T} - \frac{1}{((e^m)^T u^T)((e^n)^T v^T)} uAv^T \\ &= \frac{1}{(e^m)^T u^T} - \frac{1}{((e^m)^T u^T)((e^n)^T v^T)} vBu^T. \end{aligned}$$

If we put

$$\hat{u} = \frac{1}{(e^m)^T u^T} u \quad \text{and} \quad \hat{v} = \frac{1}{(e^n)^T v^T} v, \tag{2.58}$$

then it follows that

$$\hat{u}^T \geq \Theta_m, \ (e^m)^T \hat{u}^T = 1, \ \hat{v}^T \geq \Theta_n, \ (e^n)^T v^T = 1$$

and

$$A\hat{v}^T = \frac{1}{(e^n)^T v^T} A v^T \leq \frac{1}{(e^n)^T v^T}(e^m - kEv^T) = (\frac{1}{(e^n)^T v^T} - k)e^m = (\hat{u}A\hat{v}^T)e^m,$$
$$B\hat{u}^T = \frac{1}{(e^m)^T u^T} B u^T \leq \frac{1}{(e^m)^T u^T}(e^n - kEu^T) = (\frac{1}{(e^m)^T u^T} - k)e^n = (\hat{v}b\hat{u}^T)e^n,$$

i.e., the inequations in (2.54) hold true.
The last condition of (2.54) is also satisfied.
So in order to find a solution $(\hat{u}, \hat{v}) \in S_1 \times S_2$ one has to find a solution $(u^T, v^T) \in \mathbb{R}^m_+ \times \mathbb{R}^n_+$ with $u^T \neq \Theta_n$ and $v^T \neq \Theta_m$ of (2.57) and to define $\hat{u}$ and $\hat{v}$ by (2.58).
Let us demonstrate this by the following example:
Let $m = 3, n = 2$ and

$$A = \begin{pmatrix} -1 & 5 \\ 1 & 4 \\ 2 & 2 \end{pmatrix}, \quad B = \begin{pmatrix} 0 & -1 & 3 \\ -1 & 1 & 2 \end{pmatrix}.$$

If we choose $k = 2$, then

$$A + kE = \begin{pmatrix} 1 & 7 \\ 3 & 6 \\ 4 & 4 \end{pmatrix}, \quad B + kE^T = \begin{pmatrix} 2 & 1 & 5 \\ 1 & 3 & 4 \end{pmatrix}.$$

The initial tableaus read

		$-v_1$	$-v_2$
x_1	1	1	$\boxed{7}$
x_2	1	3	6
x_3	1	$\boxed{4}$	4

$(j_1 = 2)$

		$-u_1$	$-u_2$	$-u_3$
y_1	1	$\boxed{2}$	1	$\boxed{5}$
y_2	1	1	3	4

$(i_1 = 1)$

If we choose $j_1 = 2$ and $i_1 = 1$, i.e., we exchange in the left tableau v_2 with x_1, we obtain the tableau

		$-v_1$	$-\chi_1$
v_2	$\frac{1}{7}$	$\frac{1}{7}$	$\frac{1}{7}$
χ_2	$\frac{1}{7}$	$\boxed{\frac{15}{7}}$	$-\frac{6}{7}$
χ_3	$\frac{3}{7}$	$\frac{24}{7}$	$-\frac{4}{7}$

$(j_2 = 1)$

and we obtain $j_2 = 1$ and $i_2 = 2$. If we exchange in the right tableau u_1 with y_1, we get the tableau

		$-y_1$	$-u_2$	$-u_3$
u_1	$\frac{1}{2}$	$\frac{1}{2}$	$\frac{1}{2}$	$\frac{5}{2}$
y_2	$\frac{1}{2}$	$-\frac{1}{2}$	$\boxed{\frac{5}{2}}$	$\frac{3}{2}$

$(i_2 = 2)$

If we choose in the second left tableau $j_3 = 2 = j_1$ and $i_3 = 2 = i_2$, i.e. we exchange v_1 with χ_2, we get the tableau

		$-\chi_2$	$-\chi_1$
v_2	$\frac{2}{15}$	$-\frac{1}{15}$	$\frac{1}{5}$
v_1	$\frac{7}{15}$	$\frac{7}{15}$	$-\frac{2}{5}$
χ_3	$\frac{1}{5}$	$-\frac{8}{5}$	$\frac{4}{5}$

$(j_3 = 1 = j_1)$

If we exchange in the second right tableau u_2 with y_2, then we obtain the tableau

		$-y_1$	$-y_2$	$-u_3$
u_1	$\frac{2}{5}$	$\frac{3}{5}$	$-\frac{1}{5}$	$\frac{11}{5}$
u_2	$\frac{1}{5}$	$-\frac{1}{5}$	$\frac{2}{5}$	$\frac{3}{5}$

$(i_3 = 2 = i_2)$

The procedure now ends with the solution

$$u = (\frac{2}{5}, \frac{1}{5}, 0), \quad v = (\frac{1}{15}, \frac{2}{15})$$

of (2.55).

If we choose $j_1 = 1$ and $i_3 = 3$, i.e., we exchange in the left initial tableau v_1 with χ_3, we obtain the tableau

		$-\chi_3$	$-v_2$
χ_1	$\frac{3}{4}$	$-\frac{1}{4}$	6
χ_2	$\frac{1}{4}$	$-\frac{3}{4}$	3
v_1	$\frac{1}{4}$	$\frac{1}{4}$	1

$(j_2 = 1 = j_1)$

and we obtain $j_2 = 1 = j_1$ and $i_2 = 3 = i_1$. If we exchange in the right initial tableau u_3 with y_1, we get the tableau

		$-u_1$	$-u_2$	$-y_1$
u_3	$\frac{1}{5}$	$\frac{2}{5}$	$\frac{1}{5}$	$\frac{1}{5}$
y_2	$\frac{1}{5}$	$-\frac{3}{5}$	$\frac{11}{5}$	$-\frac{4}{5}$

$(i_2 = 3 = i_1)$

The procedure now ends with the solution

$$u = (0, 0, \frac{1}{5}), \quad v = (\frac{1}{4}, 0)$$

of (2.55).
One can also start the procedure with the right initial tableau. In this case one arrives at the two solutions

$$u = (0, \frac{1}{3}, 0), \quad v = (0, \frac{1}{7})$$

and

$$u = (0, 0, \frac{1}{5}), \quad v = (\frac{1}{4}, 0)$$

of (2.55).

References

[1] J.M. Bomze: Detecting all Evolutionarily Stable Strategies. Journal of Optimisation Theory and Applications. 75, 313-329 (1992).

[2] J. Hofbauer and K. Sigmund: Evolutionstheorie und dynamische Systeme. Verlag Paul Parey: Berlin und Hamburg 1984.

[3] W. Krabs: Spieltheorie. Dynamische Behandlung von Spielen. Verlag B.G. Teubner: Stuttgart, Leipzig, Wiesbaden 2005.

[4] J. Li: Das dynamische Verhalten diskreter Evolutionsspiele. Shaker Verlag: Aachen 1999.

[5] J. Maynard Smith: Game Theory and the Evolution of Fighting On Evolution. Edinburgh University Press. Edinburgh 1972, 8-28.

3

Four Models of Optimal Control in Medicine

3.1 Controlled Growth of Cancer Cells

In chemotherapeutic treatment of cancer one normally applies medication in periods of time with intermediate interruptions for recreation, since the stress on the body of the patient during the treatment is very high so that a recreation phase is required before the treatment is continued.

During the phases of recreation the healthy cells which are also damaged as well as the cancer cells are renewed so that every time one has to restart on a certain level of cancer cells that has to be brought to a lower level.

So the question arises whether by a permanent treatment without interruptions it is possible to achieve a final success. But before deciding for such a radical process one would like to estimate the chances of success: Since one cannot rely on experimental findings, it seems to be reasonable to make a thought experiment and to draw certain conclusions from it.

At first one needs an assumption on the uncontrolled growth of cancer. In [3] George W. Swan assumed that without therapeutic interaction the number $p = p(t)$ grows according to Gompertz's law (see Section 1.1)

$$\dot{p}(t) = \lambda p(t) ln \frac{\theta}{p(t)} \tag{3.1}$$

with an initial condition

$$p(0) = p_0 > 0.$$

This law is of the form (1.14) with $f(p(t))$ given by (1.16) where

$$\lambda_0 = \lambda(ln\theta - lnp_0) \text{ and } \gamma = \lambda. \tag{3.2}$$

In Section 1.1 we have shown that the law (3.1) describes S-shaped or logistic growth between the limits p_0 and $\theta = p_0 exp\left(\frac{\lambda_0}{\lambda}\right)$, if we assume that

$$\lambda_0 > \lambda > 0. \tag{3.3}$$

This will be done in the following.

In order to describe the effect of the medication to the growth of cancer cells it is assumed in [3] that this effect can be described in a mathematical model in the form $g(v(t)) \cdot p(t)$, where $v(t)$ is the dose of the medicament at the time t and $g(v(t))$ is the destruction rate per cancer cell and time unit. Instead of (3.1) one then obtains for the controlled growth of cancer the differential equation

$$\dot{p}(t) = \left[\lambda ln \frac{\theta}{p(t)} - g(v(t))\right] p(t). \tag{3.4}$$

About the form of the function $g = g(v)$ one can only make reasonable assumptions. If one assumes that $g(0) = 0$ and that g approaches a limit by strictly growing, then a reasonable choice can be

$$g(v) = \frac{k_1 v}{k_2 + v} \tag{3.5}$$

with positive constants k_1 and k_2. Inserting this into (3.4) leads to

$$\dot{p}(t) = \left[\lambda ln \frac{\theta}{p(t)} - \frac{k_1 v(t)}{k_2 + v(t)}\right] p(t). \tag{3.6}$$

The next question is how to measure the effect of the medication on the body. If one chooses a time interval $[0, T]$ for the treatment, then the value $\int_0^T C(t)dt$ with

$$C(t) = \text{concentration of the medicament in the body at time } t$$

could be a reasonable measure. But since $C(t)$ is unknown, in [3] the value

$$I(v) = \int_0^T v(t)dt \tag{3.7}$$

is proposed.

In order to estimate the success of the therapeutic treatment in the framework of this mathematical model we now consider the following

Problem of optimal control: For a given time $T > 0$ we prescribe a value

$$p(T) = p_T \in (0, \theta). \tag{3.8}$$

Then we look for a (continuous) control function $v = v(t), t \in [0, T]$ such that the corresponding solution $p = p(t)$ of the differential equation (3.6) with $p(0) = p_0 > 0$ satisfies the condition (3.8) and the effect $I(v)$ on the body given by (3.7) is as small as possible.

Without the end condition (3.8) the control problem obviously has the trivial solution $v \equiv 0$ which means no treatment.

If, however, a treatment takes place, then in addition we have the condition that

$$v(t) > 0 \text{ for all } t \in [0, T]. \tag{3.9}$$

This control problem now is not solved but the existence of a solution is assumed and with the aid of necessary conditions it is shown how such an optimal solution looks like.

For this purpose we at first introduce new variables and functions by the following definitions

$$\tau = \lambda t, \; y(\tau) = ln\frac{p(t)}{\theta}, \; u(\tau) = \frac{v(t)}{k_2}. \tag{3.10}$$

If we put $\delta = \frac{k_1}{\lambda}$, then the differential equation (3.6) is transferred into

$$y'(\tau) = \frac{dy}{d\tau}(\tau) = -y(\tau) - \frac{\delta u(\tau)}{1 + u(\tau)}. \tag{3.11}$$

and the corresponding initial and end conditions read

$$y(0) = y_0 = ln\frac{p_0}{\theta}, \; y(\lambda T) = y_T = ln\frac{p_T}{\theta}. \tag{3.12}$$

The control problem now consists of finding a function $u \in C[0, \lambda T]$ with

$$u(\tau) > 0 \text{ for all } \tau \in [0, \lambda T]. \tag{3.13}$$

such that (3.11) and (3.12) are satisfied and

$$J(u) = \int_0^{\lambda T} u(\tau) d\tau$$

is minimal.

The necessary conditions for optimal controls (see, e.g. [1]) imply that a solution of the problem is necessarily of the form

$$u(\tau) = Ce^{\frac{\tau}{2}} - 1, \quad \tau \in [0, \lambda T], \tag{3.14}$$

with $C > 1$. The corresponding solution $y = y(\tau)$ of (3.11) with $y = y_0$ reads

$$\begin{aligned} y(\tau) &= y_0 e^{-\tau} - \int_0^{\tau} \frac{\delta u(s)}{1 + u(s)} e^{s-\tau} ds \\ &= y_0 e^{-\tau} - \frac{\delta}{C} \int_0^{\tau} (C - e^{-\frac{s}{2}}) e^s ds e^{-\tau} \\ &= -\delta + (y_0 + \delta) e^{-\tau} + \frac{2\delta}{C} (e^{-\frac{\tau}{2}} - e^{-\tau}). \end{aligned}$$

Assumption:

$$y_0 + \delta > 0. \tag{3.15}$$

This assumption seems to be reasonable because of

$$\lim_{\tau \to \infty} y(\tau) = -\delta. \tag{3.16}$$

From (3.15) and (3.16) it then follows that

$$-\delta < y(\lambda T) < y_0, \tag{3.17}$$

if $T > 0$ is sufficiently large. In order to see this we also need that assumption (3.15) implies

$$y(\tau) > -\delta + (y_0 + \delta) e^{-\tau} > -\delta \text{ for all } \tau \in [0, \lambda T]. \tag{3.18}$$

In addition it follows that for every choice of $y_T \in (-\delta, y_0)$ the end condition $y(\lambda T) = y_T$ can be satisfied for a suitable choice of $C > 1$.

Result: Every optimal control has necessarily the form (3.14) and under the assumption (3.15) there exists a solution of (3.11) and (3.12) for $y_T \in (-\delta, y_0)$ and sufficiently large $T > 0$.

The inequalities (3.17) show that the value $y(\lambda T)$ is always larger than $-\delta$ and hence the value $p(T) = \theta e^{y(\lambda T)}$ larger than $\theta e^{-\delta}$.

So the permanent therapy with a minimal total dose of the medicament cannot make the number of cancer cells arbitrarily small.

The question now comes up to what extent this result depends on the special assumptions that were made in the mathematical model. In particular it is the question what happens when one replaces the Gompertz's law (3.1) by the general growth law (1.14) which reads

$$\dot{p}(t) = f(p(t))p(t). \tag{3.19}$$

Here we assume that the function $f : \mathbb{R}_+ \to \mathbb{R}$ is continuously differentiable with

$$f'(p) = \frac{df}{dp}(p) < 0 \text{ for all } p \in \mathbb{R}_+$$

which implies that f is strictly decreasing. Further it is assumed that

$$f(0) \in (0, \infty) \text{ and } f(p_m) = 0$$

for some $p_m > 0$ and that for every $p_0 \in (0, p_m)$ it is true that

$$\lim_{q \to p_m - 0} \int_{p_0}^{q} \frac{dq}{f(q)q} = \infty.$$

In Section 1.1 it is shown that the solution $p = p(t)$ of (3.19) with

$$p(0) = p_0 \in (0, p_m) \tag{3.20}$$

stays in $(0, p_m)$ and satisfies

$$\lim_{t \to \infty} p(t) = p_m.$$

We also replace the destruction rate (3.5) by a twice differentiable function $g : \mathbb{R}_+ \to \mathbb{R}_+$ which satisfies

$$g(0) = 0, \quad g(v) \leq k_1 \text{ for all } v \geq 0 \tag{3.21}$$

and

$$g'(v) > 0, \quad g''(v) < 0 \text{ for all } v \geq 0, \tag{3.22}$$

where $k_1 > 0$ is a given constant. The function (3.5) has all these properties.

Let us repeat now the

Problem of optimal control: Given the time $T > 0$ of therapy and the aim $p_T > 0$ with $p_T < p_0$. Find a control function $v \in C[0, T]$ with

$$v(t) > 0 \text{ for all } t \in [0, T]$$

such that the corresponding solution $p = p(t)$ of

$$\dot{p}(t) = [f(p(t)) - g(v(t))]\, p(t) \tag{3.23}$$

for $t \in (0, T)$ which satisfies the initial condition $p(0) = p_0$ also satisfies the end condition

$$p(T) = p_T \tag{3.24}$$

and minimizes

$$J(v) = \int_0^T v(t)\, dt.$$

We again assume the existence of an optimal pair $(\hat{p}, \hat{v}) \in C^1[0, T] \times C[0, T]$ and try to find information about an optimal therapy by using necessary conditions for an optimal pair. Here we use a multiplier rule as necessary condition for optimal pairs (see again [1] and Section A.1.2) from which we derive the existence of a function $\hat{\lambda} \in C^1[0, T]$ and a number $\hat{\lambda}_0 > 0$ such that

$$\dot{\hat{\lambda}}(t) = -\left[f'(\hat{p}(t))\hat{p}(t) + f(\hat{p}(t)) - g(\hat{v}(t))\right] \hat{\lambda}(t) \tag{3.25}$$

for all $t \in (0, T)$ and

$$-g'(\hat{v}(t))\hat{p}(t)\hat{\lambda}(t) = \hat{\lambda}_0 \tag{3.26}$$

for all $t \in [0, T]$.

If we define

$$\tilde{\lambda}(t) = \hat{p}(t)\hat{\lambda}(t) \text{ for } t \in [0, T],$$

then we obtain from (3.25) (using (3.23))

$$\dot{\tilde{\lambda}}(t) = -f'(\hat{p}(t))\hat{p}(t)\tilde{\lambda}(t) \text{ for all } t \in (0, T) \tag{3.27}$$

and (3.26) reads

$$-g'(\hat{v}(t))\tilde{\lambda}(t) = \hat{\lambda}_0. \tag{3.28}$$

From (3.27) we infer

$$\tilde{\lambda}(t) = \tilde{\lambda}(0)exp\left(-\int_0^t f'(\hat{p}(s))\hat{p}(s)\,ds\right), \quad t \in [0, T],$$

with

$$\tilde{\lambda}(0) = -\frac{\hat{\lambda}_0}{g'(\hat{v}(0))} < 0.$$

Further (3.28) implies

$$\begin{aligned} g'(\hat{v}(t)) &= -\frac{\hat{\lambda}_0}{\tilde{\lambda}(t)} = -\frac{\hat{\lambda}_0}{\tilde{\lambda}(0)} exp\left(\int_0^t f'(\hat{p}(s))\hat{p}(s)\,ds\right) \\ &= D\,exp\left(\int_0^t f'(\hat{p}(s))\hat{p}(s)\,ds\right) \end{aligned}$$

for all $t \in [0, T]$ where

$$D = -\frac{\hat{\lambda}_0}{\tilde{\lambda}(0)} = g'(\hat{v}(0)) > 0.$$

If $h : g'(\mathbb{R}_+) \to \mathbb{R}_+$ is the inverse function of $g' : \mathbb{R}_+ \to \mathbb{R}_+$, then it follows that

$$\hat{v}(t) = h\left(D\,exp\left(\int_0^t f'(\hat{p}(s))\hat{p}(s)\,ds\right)\right), \quad t \in [0, T]. \tag{3.29}$$

In addition it follows from (3.22) that

$$D < g'(0). \tag{3.30}$$

An optimal therapy $\hat{v} = \hat{v}(t)$ is therefore necessarily of the form (3.29).

This leads to some implications:

1. In general an optimal therapy $\hat{v} = \hat{v}(t)$ is a feedback control which is coupled with the size $\hat{p}(t)$ of the tumor at the time t unless the untreated tumor growth were such that

$$f'(\hat{p}(s))\hat{p}(s) = \delta = constant \text{ for } s \in [0, T].$$

   This is the case with Gompertz's growth law (3.1) where we have

$$f'(p)p = -\lambda \text{ for all } p \in (0, \infty)$$

   which implies that

$$\hat{v}(t) = h\left(De^{-\lambda t}\right), \qquad t \in [0, T].$$

2. Since $g'(\hat{v}(t))$ strictly decreases with increasing t (because of $f'(\hat{p}(s))\hat{p}(s) < 0$ for all $s \in [0, T]$) and likewise h is a strictly decreasing function, $\hat{v}(t)$ strictly increases with increasing t.

   If we choose in particular the destruction rate $g : \mathbb{R}_+ \to \mathbb{R}_+$ according to (3.5), then we obtain:

$$g'(v) = \frac{k_1 k_2}{(k_2 + v)^2} \text{ for } v \geq 0,$$

   hence $g'(0) = \frac{k_1}{k_2}$ and

$$h(w) = \sqrt{\frac{k_1 k_2}{w}} - k_2$$

   for all $w \in g'(\mathbb{R}_+) = (0, \frac{k_1}{k_2}]$.

   This implies in the case of Gompertz's growth law that

$$\hat{v}(t) = \sqrt{\frac{k_1 k_2}{D} e^{\lambda t}} - k_2 \text{ for all } t \in [0, T]$$

   and the condition (3.30) is equivalent to

$$\hat{v}(t) > 0 \text{ for all } t \in [0, T].$$

The question for which $T > 0$ and $p_T \in (0, p_0)$ the end condition $\hat{p}(T) = p_T$ can be satisfied cannot be answered for the general model in a similarly easy way as in the case of the Gompertz growth law and the choice of g according to (3.5). Yet for the general model it is possible to derive a necessary condition for the choice of p_T.

For that purpose we again start with the differential equation (3.23) and the initial condition $p(0) = p_0$ where $v \in C[0, T]$ with

$$v(t) > 0 \text{ for } t \in [0, T]$$

is chosen arbitrarily. We then look for some $p \in C^1[0, T]$ with

$$p(t) \in (0, p_m) \text{ for all } t \in [0, T]$$

which solves (3.23) and the initial condition $p(0) = p_0$.

Because of $g(v) \leq k_1$ for all $v \geq 0$ every such $p \in C^1[0, T]$ necessarily satisfies

$$\dot{p}(t) > -k_1 p(t) \text{ for all } t \in [0, T],$$

hence

$$p(t) > p_0 e^{-k_1 t} \text{ for all } t \in [0, T].$$

This implies for $p(T) = p_T$ that

$$p_T > p_0 e^{-k_1 T}.$$

3.2 Optimal Administration of Drugs

There are different ways by which drugs can be introduced into the body. Drugs can be injected directly into a body tissue such as a muscle, a vein or an artery. Drugs can also be given orally, then passing through the stomach and being absorbed through the walls of the intestines. In the first case the drug arrives directly at that part of the body where it is needed. This process can be modeled by one-compartment model which is presented in Section 3.2.1. In the second case a drug does not immediately arrive at its destination. This process can be modeled by a two-compartment model which is described in Section 3.2.2

3.2.1 A One-Compartment Model

We assume that the drug is given in dosages u_i at different times t_i, for $i = 0, \ldots, N-1$ with $t_0 = 0 < t_1 < \ldots < t_{N-1} < T$. After the administration of the dosage u_i it is assumed that the amount $\chi(t)$ of the drug decays exponentially within the time interval $t_i \le t \le t_{i+1}$ for $i = 0, \ldots, N-1$ (where $t_N = T$) according to the law

$$\chi(t) = \chi_i e^{-a(t-t_i)} \text{ for } t_i \le t \le t_{i+1}. \tag{3.31}$$

It is desired that the whole time interval $[0, T]$ a certain drug level χ_d is maintained. This is of course not possible. Graphically we have the following situation: Figure 3.1.

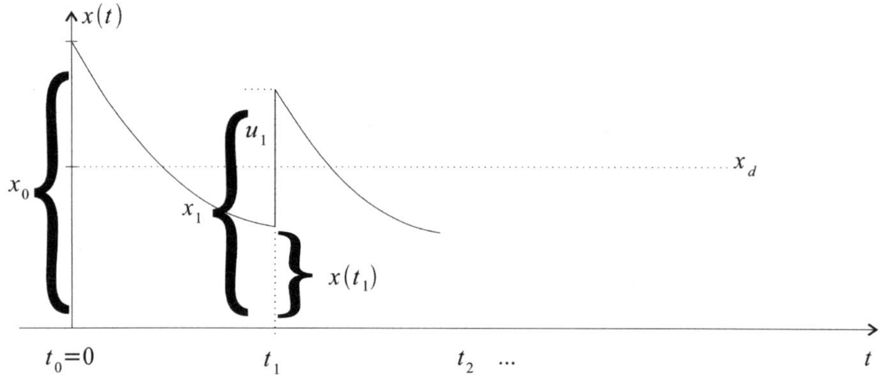

Fig. 3.1. Temporal Development of Drug Amount

This leads to the problem how to choose the dosages u_i in such a way that the integral

$$\int_0^T (\chi(t) - \chi_d)^2 dt \tag{3.32}$$

is minimized.

Because of

$$\int_0^T (\chi(t) - \chi_d)^2 dt = \sum_{i=0}^{N-1} \int_{t_i}^{t_{i+1}} (\chi(t) - \chi_d)^2 dt$$

the minimization of the integral (3.32) is equivalent to

$$\int_{t_i}^{t_{i+1}} (\chi(t) - \chi_d)^2 dt \to Min \text{ for } i = 0, \ldots, N-1 \tag{3.33}$$

where $\chi(t)$ is given by (3.31). This in turn is equivalent to

$$2\int_{t_i}^{t_{i+1}} \left(\chi_i e^{-a(t-t_i)} - \chi_d\right) e^{-a(t-t_i)} dt = 0 \text{ for } i = 0, \ldots, N-1$$

and leads to

$$\left.\begin{array}{l} \chi_i = \frac{2\chi_d}{1+exp(-a(t_{i+1}-t_i))} \\ \text{and} \\ \chi(t) = \frac{2\chi_d}{1+exp(-a(t_{i+1}-t_i))} e^{-a(t-t_i)} \text{ for } t_i \le t \le t_{i+1} \end{array}\right\} \tag{3.34}$$

and $i = 0, \ldots, N$.

From the above graphic we deduce then that

$$\begin{aligned} u_0 &= \chi_0 = \frac{2\chi_d}{1 - exp(-at_1)}, \\ u_1 &= \chi_1 - \chi(t_1) = \frac{2\chi_d}{1 - exp(-a(t_2 - t_1))} - \frac{2\chi_d}{1 + exp(-at_1)} e^{-at_1}, \\ &= \frac{2\chi_d(1 - exp(-at_2))}{(1 + exp(-a(t_2 - t_1)))(1 + exp(-at_1))}. \end{aligned}$$

In general we obtain

$$u_i = \chi_i - \chi(t_i) = \frac{2\chi_d(1 - exp(-a(t_{i+1} - t_{i-1})))}{(1 + exp(-a(t_{i+1} - t_i)))(1 + exp(-a(t_i - t_{i-1})))}$$

for $i = 1, \ldots, N-1$.

3.2.2 A Two-Compartment Model

Now we assume that the drug is administered into a first compartment in dosages a_i at times $t_i = iT$ for some $T > 0$ and $i = 0, \ldots, N-1$. After the administration of the dosage a_i it is assumed that the amount $\chi(t)$ of the drug decays exponentially within the time interval $iT \le t < (i+1)T$ for $i = 0, \ldots, N-1$ according to the law

$$\chi_1(t) = \chi_{1i} e^{-k_1(t-iT)} \text{ for } iT \le t < (i+1)T.$$

For $i = 0$ we then have

$$\chi_1(t) = a_0 e^{-k_1 t} \text{ for } 0 \le t < T.$$

This gives

$$\chi_1(T) = a_0 e^{-k_1 T} + a_1.$$

This implies

$$\chi_1(t) = \left(a_0 e^{-k_1 T} + a_1\right) e^{-k_1(t-T)} \text{ for } T \le t < 2T$$

and

$$\begin{aligned}\chi_1(2T) &= \left(a_0 e^{-k_1 T} + a_1\right) e^{-k_1 T} + a_2 = \chi_1(T) e^{-kT} + a_2 \\ &= \left(a_0 + a_1 e^{k_1 T}\right) e^{-k_1(2T)} + a_2.\end{aligned}$$

In general one obtains

$$\chi_1(t) = \left(a_0 + a_1 e^{k_1 T} + \cdots + a_i e^{k_1(iT)}\right) e^{-k_1 t}$$

for $iT \le t < (i+1)T$ and $i = 0, \ldots, N-1$.

After getting in the first compartment the drug is absorbed by the second compartment where it also decays exponentially. This process can be described by the differential equation

$$\dot{\chi}_2(t) = k_1 \chi_1(t) - k_2 \chi_2(t), \qquad t \in (0, NT), \tag{3.35}$$

with the initial condition

$$\chi_2(0) = 0 \tag{3.36}$$

where $\chi_2(t)$ denotes the amount of drug in the second compartment at the time t. We assume that $k_2 > k_1 > 0$.

The solution of (3.35), (3.36) can be given explicitly by

$$
\begin{aligned}
\chi_2(t) &= k_1 \int_0^t e^{-k_2(t-s)} \chi_1(s)\, ds \\
&= k_1 e^{-k_2 t} \int_0^t e^{k_2 s} \chi_1(s)\, ds \\
&= k_1 e^{-k_2 t} \Big\{ \chi_{10} \int_0^T e^{(k_2-k_1)s} ds + \chi_{11} \int_T^{2T} e^{(k_2-k_1)s+k_1 T} ds \\
&\quad + \cdots + \chi_{1i} \int_{iT}^t e^{(k_2-k_1)s+k_1(iT)} ds \Big\} \\
&= \frac{k_1}{k_2-k_1} e^{-k_2 t} \Big\{ \chi_{10}(e^{(k_2-k_1)T} - 1) + \chi_{11}(e^{(k_2-k_1)(2T)+k_1 T} - e^{k_2 T}) \\
&\quad + \cdots + \chi_{1i}(e^{(k_2-k_1)t+k_1(iT)} - e^{k_2(iT)}) \Big\}
\end{aligned}
$$

for $iT \le t < (i+1)T$, in particular

$$
\begin{aligned}
\chi_2(t) &= \frac{k_1}{k_2-k_1}(e^{(k_2-k_1)T} - 1)e^{-k_2 t}\chi_{10} \text{ for } 0 \le t < T, \\
\chi_2(t) &= \frac{k_1}{k_2-k_1}\left[(e^{(k_2-k_1)T} - 1)e^{-k_2 t}\chi_{10} + (e^{(k_2-k_1)(2T)+k_1 T} - e^{k_2 T})e^{-k_2 t}\chi_{11}\right] \\
&\text{for } T \le t < 2T.
\end{aligned}
$$

Again it is desired that during the whole time interval $[0, N \cdot T]$ in the second compartment a certain drug level $\alpha > 0$ is maintained. Of course this is impossible. We therefore replace this requirement by the minimization of the integral

$$
\int_0^{NT} (\chi_2(t) - \alpha)^2 dt = \sum_{i=0}^{N-1} \int_{iT}^{(i+1)T} (\chi_2(t) - \alpha)^2 dt.
$$

Let us consider the case $N = 2$ which is representative for a general $N \in \mathbb{N}$. In this case we have to minimise

$$
\begin{aligned}
f(\chi_{10}, \chi_{11}) &= \int_0^T \left(\frac{k_1}{k_2-k_1}(e^{(k_2-k_1)T} - 1)e^{-k_2 t}\chi_{10} - \alpha \right)^2 dt \\
&+ \int_T^{2T} \Big(\frac{k_1}{k_2-k_1}(e^{(k_2-k_1)T} - 1)e^{-k_2 t}\chi_{10} \\
&+ \frac{k_1}{k_2-k_1}(e^{(k_2-k_1)(2T)+k_1} - e^{k_2 T})e^{-k_2 t}\chi_{11} - \alpha \Big)^2 dt.
\end{aligned}
$$

If we put

$$A_0 = \frac{k_1}{k_2 - k_1}(e^{(k_2-k_1)T} - 1) \text{ and } A_1 = \frac{k_1}{k_2 - k_1}(e^{(k_2-k_1)(2T)+k_1 T} - e^{k_2 T}),$$

then we obtain

$$f(\chi_{10}, \chi_{11}) = \int_0^T \left(A_0 e^{-k_2 t}\chi_{10} - \alpha\right)^2 dt + \int_T^{2T} \left(A_0 e^{-k_2 t}\chi_{10} + A_1 e^{-k_2 t}\chi_{11} - \alpha\right)^2 dt.$$

Necessary and sufficient for a minimum point $(\hat{\chi}_{10}, \hat{\chi}_{11})$ is that

$$f_{\chi_{10}}(\hat{\chi}_{10}, \hat{\chi}_{11}) = f_{\chi_{11}}(\hat{\chi}_{10}, \hat{\chi}_{11}) = 0.$$

This leads to the linear system

$$\int_0^{2T} A_0^2 e^{-2k_2 t} dt \hat{\chi}_{10} + \int_T^{2T} A_1 A_0 e^{-2k_2 t} dt \hat{\chi}_{11} = \alpha \int_0^{2T} A_0 e^{-2k_2 t} dt,$$

$$\int_T^{2T} A_0 A_1 e^{-2k_2 t} dt \hat{\chi}_{10} + \int_T^{2T} A_1^2 e^{-2k_2 t} dt \hat{\chi}_{11} = \alpha \int_T^{2T} A_1 e^{-2k_2 t} dt.$$

From the unique solution $(\hat{\chi}_{10}, \hat{\chi}_{11})$ we then obtain the dosages a_0 and a_1 via

$$a_0 = \hat{\chi}_{10} \text{ and } a_0 e^{-k_1 T} + a_1 = \hat{\chi}_{11} \Rightarrow a_1 = \hat{\chi}_{11} - \hat{\chi}_{10} e^{-k_1 T}.$$

The unique solution $(\hat{\chi}_{10}, \hat{\chi}_{11})$ of the above linear system can be easily calculated and reads

$$\hat{\chi}_{10} = \frac{2\alpha}{A_0(1 + e^{-k_2 T})} \quad \text{and} \quad \hat{\chi}_{11} = \frac{2\alpha}{A_1(1 + e^{-k_2 T})}(e^{k_2 T} - 1).$$

$\hat{\chi}_{10}$ is also a minimal point of

$$\int_0^T (\chi_2(t) - \alpha)^2 dt = \int_0^T \left(\underbrace{\frac{k_1}{k_2 - k_1}(e^{(k_2-k_1)T} - 1)}_{A_0} e^{-k_2 t}\chi_{10} - \alpha \right)^2 dt$$

and $\hat{\chi}_{11}$ can be obtained by minimising

$$\int_T^{2T} \left(A_0 e^{-k_2 t} \hat{\chi}_{10} + \underbrace{\frac{k_1}{k_2 - k_1} (e^{(k_2-k_1)(2T)+k_1 T} - e^{k_2 T})}_{A_1} e^{-k_2 t} \chi_{11} - \alpha \right)^2 dt.$$

If we put

$$A_i = \frac{k_1}{k_2 - k_1} (e^{(k_2-k_1)(i+1)T + k_1(iT)} - e^{k_2(iT)}), \text{ for } i = 0, \ldots, N-1,$$

then the minimisation of

$$\int_0^{NT} (\chi_2(t) - \alpha)^2 dt$$

turns out to be equivalent to solving the *linear system*

$$\begin{aligned}
&\int_0^{NT} A_0^2 e^{-2k_2 t} dt \hat{\chi}_{10} + \int_T^{NT} A_1 A_0 e^{-2k_2 t} dt \hat{\chi}_{11} + \cdots + \int_{(N-1)T}^{NT} A_{N-1} A_0 e^{-2k_2 t} dt \hat{\chi}_{1N-1} = \alpha \int_0^{NT} A_0 e^{-k_2 t} dt, \\
&\int_T^{NT} A_0 A_1 e^{-2k_2 t} dt \hat{\chi}_{10} + \int_T^{NT} A_1^2 e^{-2k_2 t} dt \hat{\chi}_{11} + \cdots + \int_{(N-1)T}^{NT} A_{N-1} A_1 e^{-2k_2 t} dt \hat{\chi}_{1N-1} = \alpha \int_T^{NT} A_1 e^{-k_2 t} dt, \\
&\qquad \vdots \qquad\qquad\qquad\qquad \vdots \qquad\qquad\qquad\qquad \vdots \qquad\qquad\qquad\qquad \vdots \\
&\int_{(N-1)T}^{NT} A_0 A_{N-1} e^{-2k_2 t} dt \hat{\chi}_{10} + \int_{(N-1)T}^{NT} A_1 A_{N-1} e^{-2k_2 t} dt \hat{\chi}_{11} + \cdots + \int_{(N-1)T}^{NT} A_{N-1}^2 e^{-2k_2 t} dt \hat{\chi}_{1N-1} = \alpha \int_{(N-1)T}^{NT} A_{N-1} e^{-k_2 t} dt,
\end{aligned}$$

The unique solution $\hat{\chi}_{10}, \ldots, \hat{\chi}_{1N-1}$ of this system can be obtained by successive elimination of $\hat{\chi}_{1N-1}, \ldots, \hat{\chi}_{11}$ until one equation for $\hat{\chi}_{10}$ is left. $\hat{\chi}_{10}$ is than inserted into the second equation which is solved to give $\hat{\chi}_{11}$ and so forth. The dosages $a_0, a_1, \ldots, a_{N-1}$ can be calculated recursively via

$$a_0 = \hat{\chi}_{10}, a_0 e^{-k_1 T} + a_1 = \hat{\chi}_{11}, \ldots, a_0 e^{-k_1(N-1)T} + a_1 e^{-k_1(N-2)T} + \cdots + a_{N-1} = \hat{\chi}_{1N-1}.$$

Let us demonstrate this in the case $N = 3$ where the linear system is given as follows:

$$\begin{aligned}
\int_0^{3T} A_0^2 e^{-2k_2t}dt\hat{\chi}_{10} + \int_T^{3T} A_1A_0e^{-2k_2t}dt\hat{\chi}_{11} + \int_{2T}^{3T} A_2A_0e^{-2k_2t}dt\hat{\chi}_{12} &= \alpha\int_0^{3T} A_0e^{-k_2t}dt,\\
\int_T^{3T} A_1A_0e^{-2k_2t}dt\hat{\chi}_{10} + \int_T^{3T} A_1^2 e^{-2k_2t}dt\hat{\chi}_{11} + \int_{2T}^{3T} A_1A_2e^{-2k_2t}dt\hat{\chi}_{12} &= \alpha\int_T^{3T} A_1e^{-k_2t}dt,\\
\int_{2T}^{3T} A_2A_0e^{-2k_2t}dt\hat{\chi}_{10} + \int_{2T}^{3T} A_2A_1e^{-2k_2t}dt\hat{\chi}_{11} + \int_{2T}^{3T} A_2^2 e^{-2k_2t}dt\hat{\chi}_{12} &= \alpha\int_{2T}^{3T} A_2e^{-k_2t}dt.
\end{aligned}$$

Elimination of $\hat{\chi}_{12}$ leads to the *system*

$$\begin{aligned}
\int_0^{2T} A_0e^{-2k_2t}dt\hat{\chi}_{10} + \int_T^{2T} A_1e^{-2k_2t}dt\hat{\chi}_{11} \qquad\qquad &= \alpha\int_0^{2T} e^{-k_2t}dt,\\
\int_T^{2T} A_0e^{-2k_2t}dt\hat{\chi}_{10} + \int_T^{2T} A_1e^{-2k_2t}dt\hat{\chi}_{11} \qquad\qquad &= \alpha\int_T^{2T} e^{-k_2t}dt,\\
\int_{2T}^{3T} A_2A_0e^{-2k_2t}dt\hat{\chi}_{10} + \int_{2T}^{3T} A_2A_1e^{-2k_2t}dt\hat{\chi}_{11} + \int_{2T}^{3T} A_2^2 e^{-2k_2t}dt\hat{\chi}_{12} &= \alpha\int_{2T}^{3T} A_2e^{-k_2t}dt.
\end{aligned}$$

Elimination of $\hat{\chi}_{11}$ leads to the *system*

$$\begin{aligned}
\int_0^{T} A_0e^{-2k_2t}dt\hat{\chi}_{10} \qquad\qquad &= \alpha\int_0^{T} e^{-2k_2t}dt,\\
\int_T^{2T} A_0e^{-2k_2t}dt\hat{\chi}_{10} + \int_T^{2T} A_1e^{-2k_2t}dt\hat{\chi}_{11} \qquad\qquad &= \alpha\int_T^{2T} e^{-2k_2t}dt,\\
\int_{2T}^{3T} A_2A_0e^{-2k_2t}dt\hat{\chi}_{10} + \int_{2T}^{3T} A_2A_1e^{-2k_2t}dt\hat{\chi}_{11} + \int_{2T}^{3T} A_2^2 e^{-2k_2t}dt\hat{\chi}_{12} &= \alpha\int_{2T}^{3T} A_2e^{-2k_2t}dt.
\end{aligned}$$

From the first we obtain

$$\hat{\chi}_{10} = \frac{2\alpha}{A_0(1 + e^{-k_2T})}.$$

Insertion of $\hat{\chi}_{10}$ into the second equation and solving for $\hat{\chi}_{11}$ leads to

$$\hat{\chi}_{11} = \frac{2\alpha}{A_1(1 + e^{-k_2T})}(e^{k_2T} - 1).$$

Insertion of $\hat{\chi}_{10}$ and $\hat{\chi}_{11}$ into the third equation and solving for $\hat{\chi}_{12}$ leads to

$$\hat{\chi}_{12} = \frac{2\alpha}{A_2(1 + e^{-k_2T})}(1 - e^{k_2T} + e^{2k_2T}).$$

3.3 Optimal Control of Diabetes Mellitus

3.3.1 The Model

Diabetes mellitus is caused by the fact that the hormone insulin is produced in an insufficient amount in the pancreas and thereby no more sufficiently takes care of the process by which glucose is transported from the blood into the body cells or is stored in the liver in the form of glycogen. This has the effect that the glucose level in the blood gets too high which can cause a deadly coma.

Let us denote the glucose concentration in the blood at time t by $G(t)$ and the hormone concentration by $H(t)$. Then the temporal development of $G(t)$ and $H(t)$ can be described by the following differential equations:

$$\begin{aligned} \dot{G}(t) &= f_1(G(t), H(t)) + p(t), \\ \dot{H}(t) &= f_2(G(t), H(t)) \end{aligned} \tag{3.37}$$

where $p = p(t)$ denotes the increase of glucose concentration by the intake of sugar.

We assume that without the intake of sugar, i.e., $p \equiv 0$, two steady states values G_0 and H_0 will turn up with

$$f_1(G_0, H_0) = 0 \text{ and } f_2(G_0, H_0) = 0. \tag{3.38}$$

We further assume that G and H do not deviate far from G_0 and H_0 and define the difference functions

$$g(t) = G(t) - G_0, \qquad h(t) = H(t) - H_0. \tag{3.39}$$

By linearising (3.37) we obtain the differential equations

$$\begin{aligned}\dot{g}(t) &= -m_1 g(t) - m_2 h(t) + p(t),\\ \dot{h}(t) &= -m_3 h(t) + m_4 g(t)\end{aligned} \tag{3.40}$$

where

$$m_1 = -\frac{\partial f_1}{\partial G}(G_0, H_0) > 0, \qquad m_2 = -\frac{\partial f_1}{\partial H}(G_0, H_0) > 0,$$
$$m_3 = -\frac{\partial f_2}{\partial H}(G_0, H_0) > 0, \qquad m_4 = -\frac{\partial f_2}{\partial G}(G_0, H_0) > 0.$$

m_1 is positive because in case $h \equiv 0$ and $p \equiv 0$ the glucose concentration in the blood decreases by the transfer of glucose into the body cells and its storage in the liver. m_2 is positive because the presence of insulin promotes the decrease of glucose in the blood. m_3 is positive because in case $g \equiv 0$ the insulin concentration decreases as a consequence of matter transformation. m_4 is positive because in case $g > 0$ the production of the hormone in the pancreas increases with growing g.

In the case of diabetes mellitus it is assumed that no production of the hormone insulin takes place which in the above model is only possible, if $m_4 = 0$. In this case the insulin concentration decreases exponentially according to

$$h(t) = h(0)e^{m_3 t} \text{ for } t \geq 0.$$

The influence of $m_2 h(t)$ in the first equation (3.40) on the reduction of glucose in the blood decreases rapidly and the influence of the external supply of glucose described by $p(t)$ becomes dominant. In order to prevent an excess of glucose on the blood therefore an external supply of glucose is annihilated, i.e., $p \equiv 0$, and the body is given an external supply of insulin in a time dependent concentration $u = u(t)$. On putting $\chi_1 \equiv g$, $\chi_2 \equiv h$ we obtain then instead of (3.40) the equations

$$\begin{aligned}\dot{\chi}_1(t) &= -m_1 \chi_1(t) - m_2 \chi_2(t),\\ \dot{\chi}_2(t) &= -m_3 \chi_2(t) + u(t), \qquad t > 0,\end{aligned} \tag{3.41}$$

with the initial conditions

$$\chi_1(0) = g_0, \qquad \chi_2(0) = h_0 \tag{3.42}$$

where $g_0,\ h_0 > 0$ are prescribed values. The aim of the model now consists of choosing the external supply of insulin in such a way that within a given time interval $[0, T]$ a prescribed average level g_d of glucose concentration in the blood is nearly maintained and the injected amount of insulin is small. This leads to the minimisation of the integral

$$J(u) = \int_0^T (\chi_1(t,u) - g_d)^2 + \zeta u(t)^2 dt \tag{3.43}$$

by a suitable choice of $u \in C[0, T]$ where $\chi_1(\cdot, u)$ denotes the corresponding solution of (3.41), (3.42) and $\zeta > 0$ is a suitable weight factor.

3.3.2 On the Approximate Solution of the Model Problem

For a given $u \in C[0, T]$ the unique solution of the second equation (3.41) under the initial condition $\chi_2(0) = h_0$ reads

$$\chi_2(t) = e^{-m_3 t}\left\{h_0 + \int_0^t e^{m_3 s} u(s)\, ds\right\}, \quad t \in [0, T].$$

Insertion into the first equation of (3.41) leads to

$$\dot{\chi}_1(t) = -m_1\chi_1(t) - m_2 e^{-m_3 t}\left\{h_0 + \int_0^t e^{m_3 s} u(s)\, ds\right\}, \quad t \in (0, T].$$

This equation, together with the initial condition $\chi_1(0) = g_0$ has the unique solution

$$\begin{aligned}\chi_1(t) &= e^{-m_1 t}\left\{g_0 - m_2 \int_0^t e^{(m_1 - m_3)s}\left\{h_0 + \int_0^s e^{m_2 \tau} u(\tau)\, d\tau\right\} ds\right\} \\ &= e^{-m_1 t}\left\{g_0 - m_2 h_0 \int_0^t e^{(m_1 - m_3)s} ds\right\} \\ &\quad - e^{-m_1 t} m_2 \int_0^t e^{(m_1 - m_3)s} \int_0^s e^{m_3 \tau} u(\tau)\, d\tau ds.\end{aligned}$$

We assume that $m_1 \neq m_3$. If we then define

$$\chi_1^0(t) = e^{-m_1 t}\left\{g_0 - \frac{m_2 h_0}{m_1 - m_3}(e^{(m_1 - m_3)t} - 1)\right\}$$

and

$$K(t-s) = \frac{m_2}{m_1 - m_3}(e^{(m_1-m_3)(t-s)} - 1)e^{-m_1(t-s)},$$

we obtain

$$\chi_1(t) = \chi_1^0(t) - \int_0^t K(t-s)u(s)\,ds, \qquad t \in [0,T].$$

We also assume that $g_d = 0$. Then the above minimisation problem consists of finding some $u \in C[0,T]$ which minimizes

$$J(u) = \int_0^T \left(\chi_1^0(t) - \int_0^t K(t-s)u(s)\,ds\right)^2 + \zeta u(t)^2 dt.$$

In order to solve this problem approximately we replace $C[0,T]$ by a suitable n-dimensional subspace U_n which is spanned by the functions $u_1, \dots, u_n \in C[0,T]$ and consists of all linear combinations

$$u(t) = \sum_{j=1}^n \alpha_j u_j(t), \quad t \in [0,T].$$

Insertion into $J(u)$ gives

$$\begin{aligned} f(\alpha_1, \dots, \alpha_n) &= J\left(\sum_{j=1}^n \alpha_j u_j\right) \\ &= \int_0^T \left(\chi_1^0(t) - \sum_{j=1}^n w_j(t)\alpha_j\right)^2 + \zeta\left(\sum_{j=1}^n u_j(t)\alpha_j\right)^2 dt, \end{aligned}$$

if one puts

$$w_j(t) = \int_0^t K(t-s)u_j(s)\,ds \quad \text{for} \quad t \in [0,T] \quad \text{and} \quad j = 1, \dots n.$$

Instead of finding some $\hat{u} \in C[0,T]$ with

$$J(\hat{u}) \le J(u) \quad \text{for all} \quad u \in C[0,T]$$

we now look for some vector $\hat{\alpha} = (\hat{\alpha}_1, \dots, \hat{\alpha}_n)^T \in \mathbb{R}^n$ such that

$$f(\hat{\alpha}) \le f(\alpha) \quad \text{for all} \quad \alpha \in \mathbb{R}^n. \tag{3.44}$$

With this $\hat{\alpha}$ we then replace $\hat{u}$ by $\sum_{j=1}^{n} \hat{\alpha}_j u_j$.

Necessary and sufficient for $\hat{\alpha} \in \mathbb{R}^n$ to satisfy (3.44) are the conditions

$$f_{\alpha_j}(\hat{\alpha}) = 2\int_0^T \left(\chi_1^0(t) - \sum_{k=1}^{n} w_k(t)\hat{\alpha}_k\right)(-w_j(t)) + \zeta\left(\sum_{k=1}^{n} u_k(t)\hat{\alpha}_k\right) u_j(t)\, dt = 0$$

for $j = 1, \ldots, n$.

These are equivalent with the linear system

$$\sum_{k=1}^{n}\left[\int_0^T \zeta u_j(t)u_k(t) + w_j(t)w_k(t)dt\right]\hat{\alpha}_k = \int_0^T \chi_1^0(t)w_j(t)dt \quad \text{for} \quad j = 1, \ldots, n.$$

Because of the positive definiteness of the matrix

$$B = (B_{jk}) \text{ with } B_{jk} = \int_0^T \zeta u_j(t)u_k(t) + w_j(t)w_k(t)dt, \quad j,k = 1, \ldots n$$

this system has a unique solution $(\hat{\alpha}_1, \ldots, \hat{\alpha}_n)$ which gives

$$\chi_1(t) = \chi_1^0(t) - \sum_{j=1}^{n} \hat{\alpha}_j w_j(t) = \chi_1^0(t) - \int_0^T K(t-s)\hat{u}(s)\, ds$$

$$\text{where} \quad \hat{u}(t) = \sum_{j=1}^{n} \hat{\alpha}_j u_j(t).$$

Further we obtain

$$\int_0^T \left(\chi_1^0(t) - \sum_{k=1}^{n} w_k(t)\hat{\alpha}_k\right)\left(-\sum_{j=1}^{n} w_j(t)\hat{\alpha}_j\right) + \zeta\left(\sum_{k=1}^{n} u_k(t)\hat{\alpha}_k\right)^2 dt = 0$$

which implies that

$$J(\hat{u}) = \int_0^T \left(\chi_1^0(t) - \sum_{k=1}^{n} w_k(t)\hat{\alpha}_k\right)\chi_1^0(t)\, dt.$$

An Example: Let $n = 2, u_1(t) = 1, u_2(t) = t$ for $t \in [0, T]$. We choose $h_0 = 0$. Then

$$\chi_1^0(t) = e^{-m_1 t} g_0 \text{ for } t \in [0, T].$$

Further we have

$$w_1(t) = \frac{m_2}{m_1 - m_3} \int_0^t \left(e^{-m_3(t-s)} - e^{-m_1(t-s)}\right) ds$$
$$= \frac{m_2}{m_1 - m_3} \left(\frac{1}{m_3} - \frac{1}{m_1} - \frac{1}{m_3} e^{-m_3 t} + \frac{1}{m_1} e^{-m_1 t}\right)$$

and

$$w_2(t) = \frac{m_2}{m_1 - m_3} \int_0^t \left(e^{-m_3(t-s)} - e^{-m_1(t-s)}\right) s\, ds$$
$$= \frac{m_2}{m_1 - m_3} \left(\frac{t}{m_3} - \frac{t}{m_1} - \frac{1}{m_3^2}(1 - e^{-m_3 t}) + \frac{1}{m_1^2}(1 - e^{-m_1 t})\right).$$

if we put $\zeta = 1$, then the linear system

$$\left(T + \int_0^T w_1(t)^2 dt\right) \hat{\alpha}_1 + \left(\frac{T^2}{2} + \int_0^T w_1(t) w_2(t)\, dt\right) \hat{\alpha}_2 = \int_0^T e^{-m_1 t} g_0 w_1(t)\, dt,$$
$$\left(\frac{T}{2} + \int_0^T w_1(t) w_2(t)\, dt\right) \hat{\alpha}_1 + \left(\frac{T^3}{3} + \int_0^T w_2(t)^2 dt\right) \hat{\alpha}_2 = \int_0^T e^{-m_1 t} g_0 w_2(t) dt,$$

has to be solved in order to obtain the optimal control

$$\hat{u}(t) = \hat{\alpha}_1 + \hat{\alpha}_2 t, \quad t \in [0, T]$$

and the corresponding glucose concentration

$$\chi_1(t) = \chi_1^0(t) + \hat{\alpha}_1 w_1(t) + \hat{\alpha}_2 w_2(t), \quad t \in [0, T].$$

3.3.3 A Time-Discrete Diabetes Model

In order to set up a time-discrete diabetes model we introduce a time stepsize $\Delta t > 0$ such that, for a suitable $N \in \mathbb{N}$, we have $N \cdot \Delta t = T$. Then we replace in (3.41) the derivatives $\dot{\chi}_1(t)$ and $\dot{\chi}_2(t)$ by difference quotients

$$\frac{\chi_1(t + \Delta t) - \chi_1(t)}{\Delta t}$$

and

$$\frac{\chi_2(t + \Delta t) - \chi_2(t)}{\Delta t}.$$

If we define $t_k = k \cdot \Delta t$, $\chi_1^k = \chi_1(t_k)$, $\chi_2^k = \chi_2(t_k)$, and $u^k = u(t_k)$ for $k = 0, \dots, N$, then instead of (3.41) we obtain the difference equations

$$\begin{aligned}
\chi_1^{k+1} &= (1 - m_1 \Delta t)\chi_1^k - m_2 \Delta t \chi_2^k, \\
\chi_2^{k+1} &= (1 - m_3 \Delta t)\chi_2^k + \Delta t u^k \\
&\text{for } k = 0, \dots, N-1.
\end{aligned} \tag{3.45}$$

In addition we have the initial conditions

$$\chi_1^0 = g_0 \text{ and } \chi_2^0 = h_0. \tag{3.46}$$

For a given vector $(u^0, \dots, u^{N-1})$, by (3.45), (3.46) we then can compute recursively the two vectors $(\chi_1^1, \dots, \chi_1^N)$ and $(\chi_2^1, \dots, \chi_2^N)$. Finally we replace the integral (3.43) by the finite sum

$$J_N(\chi_1^1, \dots, \chi_1^N; u^0, \dots u^{N-1}) = \sum_{k=0}^{N-1} \left[(\chi_1^{k+1} - g_d^{k+1})^2 + \zeta (u^k)^2 \right] \tag{3.47}$$

where $g_k^d = g_d(t_k)$ for $k = 1, \dots, N$ are the values for a desired average glucose level at the times t_k, $k = 1, \dots, N$.

Now we are looking for a vector $(u^0, \dots, u^{N-1})$ of insulin concentrations which are given to the body at the times $t_0, \dots, t_{N-1}$ such that for the values $(\chi_1^1, \dots, \chi_1^N)$ of glucose concentrations which result from (3.45), (3.46) at the times $t_1, \dots, t_N$ the value (3.47) becomes as small as possible. For the solution of this problem we apply the Lagrangean multiplier rule.

For that purpose we rewrite the difference equation (3.45) in the form

$$\begin{aligned}
g_1(\chi_1^{k+1}, \chi_1^k, \chi_2^k) &= \chi_1^{k+1} - (1 - m_1 \Delta t)\chi_1^k - m_2 \Delta t \chi_2^k = 0, \\
g_2(\chi_2^{k+1}, \chi_2^k, u^k) &= \chi_2^{k+1} - (1 - m_3 \Delta t)\chi_2^k - \Delta t u^k = 0 \\
&\text{for } k = 0, \dots, N-1.
\end{aligned} \tag{3.48}$$

and assume, for $k = 0$, the values $\chi_1^0 = g_0$ and $\chi_2^0 = h_0$ to be inserted.

Then we define a Lagrange function by

$$
\begin{aligned}
&L(\chi_1,\chi_2,u,\lambda_1,\lambda_2)\\
&= L(\chi_1^1,\dots,\chi_1^N,\chi_2^1,\dots,\chi_2^N,u^0,\dots,u^{N-1},\lambda_1^1,\dots\lambda_1^N,\lambda_2^1,\dots\lambda_2^N)\\
&= J_N(\chi_1^1,\dots,\chi_1^N,u^0,\dots,u^{N-1},)+\sum_{k=0}^{N-1}\lambda_1^{k+1}g_1(\chi_1^{k+1},\chi_1^k,\chi_2^k)\\
&+\sum_{k=0}^{N-1}\lambda_2^{k+1}g_2(\chi_2^{k+1},\chi_2^k,u^k)
\end{aligned}
$$

where λ_1^k,λ_2^k for $k=1,\dots N$ are the so called Lagrangean multipliers.

The Lagrangean multiplier rule then reads as follows:

If, for $\hat\chi_1^{k+1},\hat\chi_2^{k+1},\hat u^k,k=0,\dots,N-1$, the conditions (3.48) are satisfied, then $J_N(\hat\chi_1^1,\dots,\hat\chi_1^N,\hat u^0,\dots,\hat u^{N-1})$ is minimal, if there exist multipliers $\hat\lambda_1^{k+1},\hat\lambda_2^{k+1},k=0,\dots,N-1$, such that

$$
\left.\begin{aligned}
&\frac{\partial L}{\partial\chi_1^{k+1}}(\hat\chi_1,\hat\chi_2,\hat u,\hat\lambda_1,\hat\lambda_2)=0,\\
&\frac{\partial L}{\partial\chi_2^{k+2}}(\hat\chi_1,\hat\chi_2,\hat u,\hat\lambda_1,\hat\lambda_2)=0,\\
&\frac{\partial L}{\partial u^k}(\hat\chi_1,\hat\chi_2,\hat u,\hat\lambda_1,\hat\lambda_2)=0,
\end{aligned}\right\}\quad\text{for}\quad k=0,\dots,N-1.
$$

In explicit form these conditions read

$$
\left.\begin{aligned}
2(\hat\chi_1^{k+1}-g_d^{k+1})+\hat\lambda_1^{k+1}-(1-m_1\Delta t)\hat\lambda_1^{k+2}&=0,\\
-m_2\Delta t\hat\lambda_1^{k+2}+\hat\lambda_2^{k+1}-(1-m_3\Delta t)\hat\lambda_2^{k+2}&=0,\\
2\zeta\hat u^k-\Delta t\hat\lambda_2^{k+1}&=0
\end{aligned}\right\}\quad\text{for}\quad k=0,\dots,N-1. \tag{3.49}
$$

where $\hat\lambda_1^{N+1}=\hat\lambda_2^{N+1}=0$.

In addition we have the conditions

$$
\begin{aligned}
\hat\chi_1^{k+1}-(1-m_1\Delta t)\hat\chi_1^k-m_2\Delta t\hat\chi_2^k&=0,\\
\hat\chi_2^{k+1}-(1-m_3\Delta t)\hat\chi_2^k-\Delta tu^k\quad&=0\\
\text{for }\ k=0,\dots,N-1.&
\end{aligned} \tag{3.50}
$$

where $\hat\chi_1^0=g_0$ and $\hat\chi_2^0=h_0$.

If one eliminates $\hat{u}^k$ from the third equation in (3.49) via

$$\hat{u}^k = \frac{1}{2\zeta}\Delta t \hat{\lambda}_2^{k+1} \tag{3.51}$$

and inserts it into the second equation of (3.50), then one obtains form (3.49) and (3.50) the linear system

$$\begin{aligned}
2(\hat{\chi}_1^{k+1} - g_d^{k+1}) + \hat{\lambda}_1^{k+1} - (1 - m_1\Delta t)\hat{\lambda}_1^{k+2} &= 0,\\
-m_2\Delta t\hat{\lambda}_1^{k+2} + \hat{\lambda}_2^{k+1} - (1 - m_3\Delta t)\hat{\lambda}_2^{k+2} &= 0,\\
\hat{\chi}_1^{k+1} - (1 - m_1\Delta t)\hat{\chi}_1^k - m_2\Delta t\hat{\chi}_2^k &= 0,\\
\hat{\chi}_2^{k+1} - (1 - m_3\Delta t)\hat{\chi}_2^k + \frac{(\Delta t)^2}{2\zeta}\hat{\lambda}_2^{k+1} \quad &= 0\\
\text{for } k = 0, \ldots, N-1.
\end{aligned} \tag{3.52}$$

where $\hat{\chi}_1^0 = g_0$, $\hat{\chi}_2^0 = h_0$ and $\hat{\lambda}_1^{N+1} = \hat{\lambda}_2^{N+1} = 0$.

In order to solve the above problem one then has to solve the linear system (3.52) for $\hat{\chi}_1^{k+1}, \hat{\chi}_2^{k+1}, \hat{\lambda}_1^{k+1}, \hat{\lambda}_1^{k+1}$, for $k = 0, \ldots N-1$ and define $\hat{u}^k$ for $k = 0, \ldots N-1$ by (3.51).

In [2] it is shown how the linear system (3.52) can be solved by a shooting method.

3.3.4 An Exact Solution of the Model Problem

The minimization of J on $C[0,T]$ can be reduced to the solution of a Fredholm integral equation on $C[0,T]$ which is uniquely solvable with the aid of successive approximation, if ζ is large enough. In order to show that we first observe that J is Fréchet differentiable on $C[0,T]$ and the Fréchet derivative $J_u' : C[0,T] \to C[0,T]$ is given by

$$J_u'(h) = 2\int_0^T \left[\chi_1^0(t) + \int_0^t K(t-s)u(s)\,ds\right]\int_0^t K(t-s)h(s)\,ds + \zeta u(t)h(t)\,dt$$
$$\text{for every } h \in C[0,T] \text{ and } u \in C[0,T].$$

If we define

$$L(t,s)=\begin{cases}K(t-s) & \text{for } 0\le s\le t,\\ 0 & \text{for } t<s\le T\end{cases}$$

and put

$$\begin{aligned}v(t)&=\chi_1^0(t)+\int_0^t K(t-s)u(s)\,ds\\ &=\chi_1^0(t)+\int_0^T L(t,s)u(s)\,ds,\end{aligned}$$

then it follows that

$$\begin{aligned}\int_0^T v(t)\int_0^t K(t-s)h(s)\,ds&=\int_0^T v(t)\int_0^T L(t,s)h(s)\,ds\,dt\\ &=\int_0^T h(s)\int_0^T L(t,s)v(t)\,dt\,ds\end{aligned}$$

and we obtain

$$\begin{aligned}J_u'(h)&=2\int_0^T h(s)\left[\int_0^T L(t,s)\left(\chi_1^0(t)+\int_0^T L(t,s)u(s)\,ds\right)\,dt+\zeta u(s)\right]ds\\ &=2\int_0^T h(s)\left[\int_0^T L(t,s)\left(\chi_1^0(t)+\tilde{L}(u)(t)\right)\,dt+\zeta u(s)\right]ds,\end{aligned}$$

if we put

$$\tilde{L}(u)(t)=\int_0^T L(t,s)u(s)\,ds,\quad t\in[0,T].$$

Since J is strictly convex on $C[0,T]$, $\hat{u}\in C[0,T]$ satisfies

$$J(\hat{u})\le J(u)\quad\text{for all } u\in C[0,T],$$

if and only if

$$J'_{\hat{u}}(h) = 0 \quad \text{for all } h \in C[0, T]$$

which is equivalent with

$$\int_0^T L(t, s)\left(\chi_1^0 + \tilde{L}(\hat{u})(t)\right) dt + \zeta \hat{u}(s) = 0 \quad \text{for all } s \in [0, T].$$

Now we have

$$\begin{aligned}
\int_0^T L(t, s)\tilde{L}(\hat{u})(t)dt &= \int_0^T L(t, s) \int_0^T L(t, \tau)u(\tau)d\tau \, dt \\
&= \int_0^T \int_0^T L(t, s)L(t, \tau) \, dt \, u(\tau) \, d\tau \\
&= \int_0^T \tilde{K}(s, \tau)u(\tau) \, d\tau,
\end{aligned}$$

if we put

$$\tilde{K}(s, \tau) = \int_0^T L(t, s)L(t, \tau) \, dt, \quad s, \tau \in [0, T].$$

With

$$f(s) = -\int_0^T L(t, s)\chi_1^0(t) \, dt, \quad s \in [0, T],$$

we then obtain the Fredholm integral equation

$$\zeta \hat{u}(s) + \int_0^T \tilde{K}(s, \tau)\hat{u}(\tau) \, d\tau = f(s), \quad s \in [0, T]$$

as a necessary and sufficient condition for $\hat{u} \in C[0, T]$ being a minimizer of J on $C[0, T]$. This equation has a unique solution, if

$$\frac{1}{\zeta} \max_{s \in [0,T]} \int_0^T \left|\tilde{K}(s, \tau)\right| d\tau < 1,$$

and this solution can be obtained by the method of successive approximation.

3.4 Optimal Control Aspects of the Blood Circulation in the Heart

3.4.1 Blood Circulation in the Heart

In this section we give a verbal description of the blood circulation in the heart. Venous blood enters the right atrium and passes in the right ventricle. During a contraction of the heart by which the valve that separates the right atrium and the right ventricle is closed the blood is pushed into the pulmonary artery. The pulmonary artery branches to the right and left lungs where the blood is oxygenated and carbon dioxide is extracted. Then the blood returns through the pulmonary veins into the left half of the heart and flows from the left atrium into the left ventricle. Contraction of the heart pushes the blood into the aorta and from there into the rest of the arterial and venous system.

The periodic contractions of the heart result in a pulsatic flow of blood into the aorta. More specifically, as a consequence of the contraction of the left ventricle there is a "pressure wave" and a "flow wave" through the vascular system. The pressure climbs rapidly to its greatest (systolic) level of about 120mm Hg. During the relaxation phase of the left heart the pressure in the left ventricle falls below the pressure in the aorta and so causes the left valve between the left ventricle and the aorta to close. This results in a decrease in the aortic pressure to its lowest (distolic) point of about 80mm Hg.

3.4.2 A Model of the Left-Ventricular Ejection Dynamics

The dynamic equations of the model are as follows:

$$V(t) = V_0 - \int_0^t i(s)\, ds, \tag{3.53}$$

$$P(t) = P_a(t) + ri(t) + L\frac{di}{dt}(t), \tag{3.54}$$

$$i(t) = \frac{1}{R}P_a(t) + C\frac{dP_a}{dt}(t) \tag{3.55}$$

where $P(t)$ and $V(t)$, respectively, represent the instantaneous values of left-ventricle pressure and volume, respectively, $P_a(t)$ denotes the blood pressure in the aorta, $i(t)$ the rate at which the blood flow is ejected out of the ventricle, r the aortic valvular resistance, L the inertia of the blood, and R and C the peripheral resistance and compliance of a lumped arterial Windkissel load.

The quantity V_0 in (3.53) represents the ventricular volume at the beginning of ejection, and the second term (3.53) is the blood volume ejected out of the ventricle during the time duration from zero to t. The interpretation of (3.54) is that the ventricular pressure is equal to the sum of aortic pressure, the pressure drop across the aortic valve (which is proportional to the blood flow rate $i(t)$), and the pressure accelerating the blood (which is proportional to $\frac{di}{dt}(t)$). Equation (3.55) is a mathematical statement of the fact that the blood flow rate into the arterical system is equal to the blood flow rate leaking out of it (which is proportional to $P_a(t)$) together with the rate of blood storage in the system (which is proportional to $\frac{dP_a}{dt}(t)$).

The model described by the three equations (3.53), (3.54), (3.55) is the analog of an electrical circuit (see [3]). Its purpose is to determine the so called elastance function of the left ventricle which is defined by $E(t) = \frac{P(t)}{V(t)}$ and can be considered qualitatively as the time-varying stiffness of the ventricle. For the determination of $E(t)$ optimal control theory is to be applied. As control variable $P(t)$ is chosen and the following performance criterion is considered

$$J = \int_0^{t_f} \left[P(t)^2 + \alpha P(t) i(t)\right] dt \tag{3.56}$$

where α has the dimension of $\frac{mmHg}{ml/sec}$. By t_f the duration of the ejection period is denoted. In addition to the equations (3.53) - (3.55) boundary conditions at $t = 0$ and $t = t_f$ are prescribed by the form

$$V(0) = V_0, \quad V(t_f) = V_f, \tag{3.57}$$

$$i(0) = i(t_f) = 0, \tag{3.58}$$

$$P_a(0) + P_a(t_f) = 2\overline{P}_a. \tag{3.59}$$

In the second equation of (3.57) the quantity V_f is given by $V_f = V_0 - V_s$ where the stroke volume V_s is obtained from the expression

$$V_s = b + \left(c - \bar{P}_a d\right) V_0$$

with given constants b, c and d. The blood flow is assumed to be zero at the beginning and at the end of the ejection which leads to the condition (3.58). The quantitiy $\bar{P}_a$ is the average aortic pressure during the ejection and is approximated by the expression $\frac{[P_a(0)+P_a(t_f)]}{2}$.

3.4.3 An Optimal Control Problem

Let us rename the variables V, i, P_a and P by

$$\chi_1 = V, \quad \chi_2 = i, \quad \chi_3 = P_a \text{ and } u = P.$$

Then the equations (3.53) - (3.55) can be rewritten in the form

$$\dot{\chi}_1(t) = -\chi_2(t), \tag{3.60}$$

$$\dot{\chi}_2(t) = \frac{u(t) - \chi_3(t) - r\chi_2(t)}{L}, \tag{3.61}$$

$$\dot{\chi}_3(t) = -\frac{1}{RC}\chi_3(t) + \frac{1}{C}\chi_2(t). \tag{3.62}$$

The performance criterion (3.56) reads

$$J(u) = \int_0^{t_f} \left(u(t)^2 + \alpha u(t)\chi_2(t)\right) dt \tag{3.63}$$

with $\alpha > 0$.
The boundary conditions (3.57) - (3.59) are reformulated as

$$\chi_1(0) = V_0, \quad \chi_1(t_f) = V_f, \tag{3.64}$$

$$\chi_2(0) = \chi_2(t_f) = 0, \tag{3.65}$$

$$\chi_3(0) + \chi_3(t_f) = 2\bar{P}_a. \tag{3.66}$$

The problem to be solved now consists of finding a control function $u \in C[0, t_f]$ such that unter the conditions (3.60) - (3.62) and (3.64) - (3.66) the functional $J = J(u)$ (3.63) is minimized.
In this form it is posed in [3] and solved with the aid of Pontryagin's minimum principle. Here we give a different solution. Since $J(u)$ does not depend on χ_1, we minimize $J(u)$, $u \in [0, t_f]$, subject to (3.61), (3.62), (3.65) and

$$\chi_3(0) = P_a(0), \quad \chi_3(t_f) = P_a(t_f) \tag{3.67}$$

instead of (3.66). We rewrite the system (3.61), (3.62) in the form

$$\begin{pmatrix} \dot{\chi}_2(t) \\ \dot{\chi}_3(t) \end{pmatrix} = A \begin{pmatrix} \chi_2(t) \\ \chi_3(t) \end{pmatrix} + bu(t), \quad t \in (0, t_f), \tag{3.68}$$

where

$$A = \begin{pmatrix} -\frac{r}{L} & -\frac{1}{L} \\ \frac{1}{C} & -\frac{1}{RC} \end{pmatrix} \quad \text{and} \quad b = \begin{pmatrix} \frac{1}{L} \\ 0 \end{pmatrix}.$$

We assume that A has two different real eigenvalues λ_1, λ_2 which is the case when

$$\left(\frac{r}{L} - \frac{1}{RC}\right)^2 - \frac{1}{LC} > 0.$$

Then A has two real eigenvectors $y_1, y_2 \in \mathbb{R}^2$ which are linearly independent (in fact infinitely many such pairs) and

$$\bar{Y}(t) = \left(y_1 e^{\lambda_1 t} | y_2 e^{\lambda_2 t}\right)$$

is a fundamental matrix function of the corresponding homogenous system.

The solution of (3.68) and the initial conditions

$$\chi_2(0) = 0 \text{ and } \chi_3(0) = P_a(0)$$

is therefore given by

$$\begin{pmatrix} \chi_2(t) \\ \chi_3(t) \end{pmatrix} = \bar{Y}(t) \left(\bar{Y}(0)^{-1} \begin{pmatrix} 0 \\ P_a(0) \end{pmatrix} + \int_0^t \bar{Y}(s)^{-1} bu(s)\, ds \right), \quad t \in [0, t_f].$$

The second equations in (3.65), (3.67) lead to the condition

$$\int_0^{t_f} \overline{Y}(t)^{-1} bu(t)dt = \overline{Y}(t_f)^{-1} \begin{pmatrix} 0 \\ P_a(t_f) \end{pmatrix} - \overline{Y}(0)^{-1} \begin{pmatrix} 0 \\ P_a(0) \end{pmatrix}.$$

If we put

$$B(t) = \overline{Y}(t)^{-1} b \text{ and } g = \overline{Y}(t_f)^{-1} \begin{pmatrix} 0 \\ P_a(t_f) \end{pmatrix} - \overline{Y}(0)^{-1} \begin{pmatrix} 0 \\ P_a(0) \end{pmatrix},$$

then this condition reads

$$\int_0^{t_f} B(t)u(t)\, dt = g. \tag{3.69}$$

If we define

$$G(t) = \overline{Y}(t)\overline{Y}(0)^{-1} \begin{pmatrix} 0 \\ P_a(0) \end{pmatrix} \text{ and } C(t, s) = \overline{Y}(t)B(s) \quad \text{for } 0 \le s \le t \le t_f, \tag{3.70}$$

then we get

$$\chi_2(t) = G_1(t) + \int_0^t C_1(t, s)u(s)\, ds.$$

The problem we have to solve now consists of finding some $u \in C[0, t_f]$ which satisfies the condition (3.69) and minimizes

$$J(u) = \int_0^{t_f} u(t)^2 + \alpha \left(G_1(t) + \int_0^t C_1(t, s)u(s)\, ds \right) u(t)\, dt. \tag{3.71}$$

From (3.70) one can infer that $C_1(t, s)$ can be represented in the form

$$C_1(t, s) = D_1 e^{\lambda_1 (t-s)} + D_2 e^{\lambda_2 (t-s)}. \tag{3.72}$$

Now let $\hat{u} \in C[0, t_f]$ be a solution of the problem.

Then one can show with the aid of a multiplier rule (see [1]) that there exist multipliers $l_1, l_2 \in \mathbb{R}$ such that

$$\hat{u}(t) - \frac{\alpha}{2} \int_0^{t_f} K(t,s)\hat{u}(s)\, ds = -\frac{\alpha}{2} G_1(t) - l_1 B_1(t) - l_2 B_2(t), \quad t \in [0, t_f], \tag{3.73}$$

where

$$K(t,s) = \begin{cases} -C_1(t,s) & \text{for } 0 \le s \le t, \\ -C_1(s,t) & \text{for } t \le s \le t_f \end{cases}.$$

From (3.72) we induce that

$$K(t,s) = \begin{cases} -D_1 e^{\lambda_1 t} e^{-\lambda_1 s} - D_2 e^{\lambda_2 t} e^{-\lambda_2 s} & \text{for } 0 \le s \le t, \\ -D_1 e^{\lambda_1 s} e^{-\lambda_1 t} - D_2 e^{\lambda_2 s} e^{-\lambda_2 t} & \text{for } t \le s \le t_f \end{cases}.$$

If we define

$$\begin{aligned} a_i(t) &= -D_i e^{\lambda_i t} \quad \text{and } b_i(s) = e^{\lambda_i s} \quad \text{for } 0 \le s \le t, \\ a_i(t) &= -D_i e^{\lambda_i s} \quad \text{and } b_i(s) = e^{\lambda_i t} \quad \text{for } t \le s \le t_f,\ i = 1, 2, \end{aligned}$$

then we can write

$$K(t,s) = a_1(t)b_1(s) + a_2(t)b_2(s) \quad \text{for } 0 \le s, t \le t_f.$$

On defining

$$A_k = \int_0^{t_f} b_k(s)\hat{u}(s)\, ds \quad \text{and } h(t) = -\frac{\alpha}{2} G_1(t) - l_1 B_1(t) - l_2 B_2(t),\ t \in [0, t_f],$$

one can rewrite (3.73) in the form

$$\hat{u}(t) = h(t) + \frac{\alpha}{2} \left(A_1 a_1(t) + A_2 a_2(t) \right), \quad t \in [0, t_f]. \tag{3.74}$$

In order to determine A_1 and A_2 we define

$$\int_0^{t_f} a_m(s)b_k(s)\,ds = \alpha_{km}, \quad k,m = 1,2,$$
$$\int_0^{t_f} h(s)b_k(s)\,ds = h_k, \quad k = 1,2.$$

Inserting (3.74) into (3.73) then leads to the linear system

$$A_k - \frac{\alpha}{2}\sum_{m=1}^{2} \alpha_{km}A_m = h_k, \quad k = 1,2. \tag{3.75}$$

This has a unique solution (A_1, A_2), if $\alpha > 0$ is chosen small enough. In order to determine l_1 and l_2 we insert (3.73) into (3.69) and obtain the linear system

$$\begin{aligned}
&\int_0^{t_f} B_1(t)^2 dt\; l_1 + \int_0^{t_f} B_1(t)B_2(t)\,dt\; l_2 \\
&\quad = g_1 + \alpha \int_0^{t_f} B_1(t)\Big(G_1(t) + \frac{1}{2}\left(A_1a_1(t) + A_2a_2(t)\right)\Big)\,dt, \\
&\int_0^{t_f} B_2(t)B_1(t)\,dt\; l_1 + \int_0^{t_f} B_2(t)^2 dt\; l_2 \\
&\quad = g_2 + \alpha \int_0^{t_f} B_2(t)\Big(G_1(t) + \frac{1}{2}\left(A_1a_1(t) + A_2a_2(t)\right)\Big)\,dt.
\end{aligned} \tag{3.76}$$

Result: In order to solve the above optimization problem one has to solve the linear system (3.75) and then the linear system (3.76) and to define $\hat{u} = \hat{u}(t)$ by (3.74).

The left-ventricle pressure and volume, respectively, are then given by

$$P(t) = \hat{u}(t)$$

and

$$V(t) = V_0 - \int_0^t \chi_2(s)\,ds, \quad t \in [0, t_f].$$

3.4.4 Another Model of the Left-Ventricular Ejection Dynamics

This model is also an optimal control model (see [3]) which is given as follows:
Under the condition

$$\dot{\chi}(t) + \frac{1}{\tau}\chi(t) = u(t), \quad t \in [0, t_s], \tag{3.77}$$

$$\chi(0) = \chi_0, \tag{3.78}$$

$$\int_0^{t_s} u(t)\, dt = V_s \quad (u \in C[0, t_s]) \tag{3.79}$$

the functional

$$J(u) = \int_0^{t_s} \left(R_C u(t)^2 + R_P\, \chi(u,t)^2\right) dt \tag{3.80}$$

is to be minimized

Here $\chi = \chi(t)$ is the peripheral flow, $u = u(t)$ is the aortic flow (which is the control variable), $\tau = R_P C_A$ where R_P is the peripheral resistance, C_A the arterial compliance, $[0, t_s]$ is the time interval from the beginning of ejection to the highest pressure (systolic), V_S is the (constant) stroke volume, and R_C the valvular resistance.
This model has also an electric analog (see [3]).
The unique solution $\chi = \chi(t), t \in [0, t_s]$, of (3.77), (3.78) can be represented in the form

$$\chi(t) = e^{-\lambda t}\left(\chi_0 + \int_0^t e^{\lambda s} u(s)\, ds\right) = e^{-\lambda t}\chi_0 + \int_0^t e^{\lambda(s-t)} u(s)\, ds$$

for $t \in [0, t_s]$ where $\lambda = \frac{1}{\tau}$.

So we have to solve the following problem:
Find some $u \in C[0, t_s]$ which satisfies (3.79) and minimizes the functional

$$J(u) = \int_0^{t_s} \left(R_C u(t)^2 + R_P \left(e^{-\lambda t}\chi_0 + \int_0^t e^{\lambda(s-t)} u(s)\, ds\right)^2\right) dt.$$

Let $\hat{u} \in C[0, t_s]$ be a solution of this problem.

Then one can show with the aid of a multiplier rule (see [1]) that there exists a multiplier $l \in \mathbb{R}$ such that

$$\int_0^{t_s} \left(2R_C\hat{u}(t) + 2R_P\left[(t_s - t)e^{\lambda t}\chi_0 + \int_0^{t_s} e^{-2\lambda s} \int_0^{s} e^{\lambda\tau}\hat{u}(\tau)d\tau ds e^{\lambda t}\right] + l\right)h(t)dt$$
$$= 0 \quad \text{for all} \quad h \in C[0, t_s].$$

This can be shown to be equivalent to the integral equation

$$\hat{u}(t) + \frac{1}{2\lambda}\frac{R_P}{R_C}\int_0^{t_s} K(t,\tau)\hat{u}(\tau)\,d\tau e^{\lambda t} = -\frac{l}{2R_C} - \frac{R_P}{R_C}\chi_0(t_s - t)e^{\lambda t}, \quad t \in [0, t_s],$$

where

$$K(t,\tau) = \begin{cases} (e^{-2\lambda t} - e^{-2\lambda t_s})e^{\lambda\tau} & \text{for } 0 \le \tau \le t, \\ (e^{-2\lambda\tau} - e^{-2\lambda t_s})e^{\lambda\tau} & \text{for } t \le \tau \le t_s. \end{cases}$$

If we put

$$\rho = \frac{1}{2\lambda}\frac{R_P}{R_C} \quad \text{and define} \quad f(t) = -\frac{l}{2R_C} - \frac{R_P}{R_C}\chi_0(t_s - t)e^{\lambda t}, \quad t \in [0, t_s], \tag{3.81}$$

then the integral equation takes the form

$$\hat{u}(t) + \rho\int_0^{t_s} K(t,\tau)\hat{u}(\tau)d\tau e^{\lambda t} = f(t), \quad t \in [0, t_s]. \tag{3.82}$$

If we put

$$A = \int_0^{t_s} K(t,\tau)\hat{u}(\tau)d\tau,$$

then we obtain

$$\hat{u} = f(t) - \rho A e^{\lambda t}, \quad t \in [0, t_s].$$

Insertion into (3.82) and solving for A leads to

$$A = \frac{\int_0^{t_s} K(t,\tau) f(\tau) d\tau}{e^{\lambda t} + \rho \int_0^{t_s} K(t,\tau) e^{\lambda \tau} d\tau},$$

hence,

$$\hat{u}(t) = f(t) - \rho \frac{\int_0^{t_s} K(t,\tau) f(\tau) d\tau}{e^{\lambda t} + \rho \int_0^{t_s} K(t,\tau) e^{\lambda \tau} d\tau}, \qquad t \in [0, t_s],$$

where ρ and $f(t)$ are given by (3.81).
The multiplier $l \in \mathbb{R}$ can then be determinded by the condition (3.79).

References

[1] A.D. Ioffe und V.M. Tichomirov: Theorie der Extremalaufgaben. VEB Deutscher Verlag der Wissenschaften: Berlin 1979.

[2] W. Krabs: Mathematische Modellierung. Verlag B.G. Teubner: Stuttgart 1997.

[3] G.W. Swan: Applications of Optimal Control Theory in Biomedicine. Marcel Dekker Inc.: New York and Basel, 1984.

4

A Mathematical Model of Hemodialysis

4.1 A One-Compartment Model

4.1.1 The Mass Transport in the Dialyzer

In most of the common dialyzers (called artificial kidneys) the removal of toxical substances from the blood is achieved by extracting it from the body and introducing it into the interior of a kidney machine where it floats along one side of a membrane as being schematically shown in Figure 4.1.

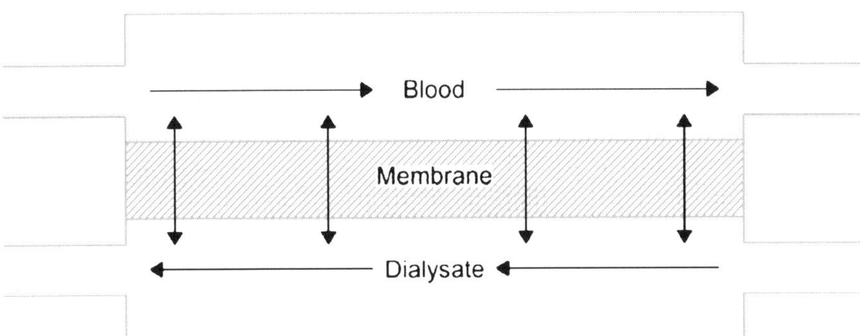

Fig. 4.1. Blood and Dialysate Flow in the Dialyzer.

At the same time a dialysate fluid floats along the other side of the membrane in the opposite direction. In the course of this process a mass transport of the toxical substance across the membrane takes place which is based on two mechanisms. The first is diffusion by which the molecules of the substance are pressed through the pores of the membrane under the influence of the difference $C_B - C_D$ of the concentration C_B $[mg/ml]$ and C_D $[mg/ml]$ of the toxical substance in the blood and the dialysate fluid, respectively. The flux F_D $[mg/min]$ of the substance across the membrane from the blood into the dialysate fluid caused by diffusion is given, according to Fick's law, by

$$F_D = D_M(C_B - C_D) \tag{4.1}$$

where D_M $[ml/min]$ is the diffusive clearance of the membrane which can be expressed by

$$D_M = K_M \cdot A$$

where $A[cm^2]$ is the area of the membrane and $K_M[cm/min]$ is the mass transfer coefficient (this law holds for thin membranes).

The second mechanism of mass transport across the membrane is called ultrafiltration and is caused by a pressure difference between the blood water and the dialysate fluid. This pressure difference gives rise to a blood water flow through the membrane by which dissolved parts of the toxical substance are transported.

Let Q_F $[ml/min]$ and Q_B $[ml/min]$ and Q_D $[ml/min]$ be the flow rate of the blood water through the membrane and the flow rate of the blood and the flow rate of the dialysate fluid (the latter two through the dialyzer), respectively. Further let $C_{B_i} = C_B$ $[mg/ml]$ and C_{B_0} $[mg/ml]$ be the concentration of the toxical substance in the blood entering the dialyzer and leaving it or passing the membrane (in the blood water), respectively. The overall flux F_M $[mg/min]$ of the toxical substance through the membrane from the blood to the dialysate fluid is given by

$$F_M = Q_B(C_{B_i} - C_{B_0}) + Q_F C_{B_0} \tag{4.2}$$

where $Q_B(C_{B_i} - C_{B_0}) = Q_B(C_B - C_{B_0})$ is the diffusion part which is equal to F_D given by (4.1), hence

$$Q_B(C_B - C_{B_0}) = D_M(C_B - C_D) \tag{4.3}$$

and $Q_F C_{B_0}$ is the part caused by ultrafiltration.

Solving (4.3) for C_{B_0} leads to

$$C_{B_0} = C_B - \frac{D_M}{Q_B}(C_B - C_D) \tag{4.4}$$

and inserting into

$$F_M = D_M(C_B - C_D) + Q_F C_{B_0}$$

yields

$$\begin{aligned} F_M &= D_M(C_B - C_D) + Q_F\left(C_B - \frac{D_M}{Q_B}(C_B - C_D)\right) \\ &= \left(D_M + Q_F\left(1 - \frac{D_M}{Q_B}\right)\right)C_B - D_M\left(1 - \frac{Q_F}{Q_B}\right)C_D. \end{aligned} \tag{4.5}$$

Realistic values for urea (as toxical substance) are (see [5], p. 50)

$$D_M = 150\ [ml/min] \qquad \text{and} \qquad Q_F = 10\ [ml/min].$$

The flow rate of blood through the dialyzer can be assumed to be $Q_B = 200\ [ml/min]$ as a reasonable average value. From these values we obtain

$$Q_F\left(1 - \frac{D_M}{Q_B}\right) = 2.5$$

and

$$\frac{Q_F}{Q_B} = 0.05.$$

In view of (4.5) we see that in the case of urea ultrafiltration can be practically neglected which amounts to putting $Q_F = 0$ and leads to

$$F_M = F_D = D_M(C_B - C_D). \tag{4.6}$$

In the case of the so called middle molecules, however, like vitamin B_{12} ultrafiltration cannot be neglected. In this case diffusion is limited and ultrafiltration can be a comparable part of mass transport through the membrane augmenting the diffusive clearance by as much as 45% (see [5], p. 45).

4.1.2 The Temporal Development of the Toxin Concentration in the Blood without Ultrafiltration

Let $V_B(t)$ $[ml]$ be the total blood volume of the patient at the time t. Then the total amount of the toxical substance in the blood at the time t is given by $V_B(t) \cdot C_B(t)$ where $C_B(t)$ $[mg/ml]$ denotes the concentration of the toxical substance at the time t. We assume that some rest clearance D_r $[ml/min]$

of the kidneys of the patient is present such that there is a permanent flux F_r $[mg/min]$ of the toxical substance out of the blood across the kidneys into the urine which is of the form

$$F_r(t) = D_r C_B(t). \tag{4.7}$$

We further assume that, within a given time interval $[0, T]$ the patient is attached to a dialyzer during the time interval $[0, t_d]$ where $t_d < T$. Then the total flux of the toxical substance out of the blood during the time period $[0, T]$ is given by (see (4.5))

$$F(t) = \begin{cases} \left(D_M + Q_F\left(1 - \frac{D_M}{Q_B}\right) + D_r\right) C_B(t) - D_M\left(1 - \frac{Q_F}{Q_B}\right) C_D(t) & \text{for } t \in [0, t_d), \\ D_r C_B(t) & \text{for } t \in [t_d, T) \end{cases} \tag{4.8}$$

where $C_D(t)$ $[mg/ml]$ denotes the concentration of the toxical substance in the dialysate fluid at the time t.

The temporal change of the concentration of the toxical substance is governed by the differential equation

$$\frac{d}{dt}(V_B(t) C_B(t)) = -F(t) + G \tag{4.9}$$

where $F(t)$ is given by (4.8) and G $[mg/min]$ is the average generation rate of the toxical substance.

If we assume $C_D(t)$ to be known for $t \in [0, T]$, then the two unknown functions in (4.9) are $V_B(t)$ and $C_B(t)$.

In the sequel we will be mainly interested in the hemodialysis of urea. Therefore we will assume henceforth that ultrafiltration can be neglected, i.e. $Q_F = 0$ (as being shown in Section 4.1.1). This implies that the blood volume $V_B(t)$ does no more depend on the time t and can be assumed to be a constant V_B $[ml]$. Therefore (4.9) takes the form

$$V_B \dot{C}_B(t) = \left\{ \begin{array}{ll} -(D_M + D_r) C_B(t) + D_M C_D(t) + G & \text{for } 0 \leq t < t_d, \\ -D_r C_B(t) + G & \text{for } t_d \leq t < T, \end{array} \right\} \tag{4.10}$$

where $\dot{C}_B$ denotes the derivative with respect to t.

In [5], p. 43 it is stated that in most single pass dialyzers and for normal catabolites the concentration $C_D(t)$ of the toxical substance in the dialysate fluid is small compared to $C_B(t)$ and can therefore be neglected. Consequently we assume

$$C_D(t) = 0 \qquad \text{for} \qquad t \in [0, t_d].$$

The solution of (4.10) for $D_r > 0$ is then given by

$$C_B(t) = C_B(0)e^{-\frac{D_M+D_r}{V_B}t} + \frac{G}{D_M + D_r}\left(1 - e^{-\frac{D_M+D_r}{V_B}t}\right) \tag{4.11}$$

for $t \in [0, t_d]$ and by

$$C_B(t) = C_B(t_d)e^{-\frac{D_r}{V_B}(t-t_d)} + \frac{G}{D_r}\left(1 - e^{-\frac{D_r}{V_B}(t-t_d)}\right) \tag{4.12}$$

for $t \in (t_d, T]$.

If $D_r = 0$, then the solution of (4.10) is given by

$$C_B(t) = C_B(0)e^{-\frac{D_M}{V_B}t} + \frac{G}{D_M}\left(1 - e^{-\frac{D_M}{V_B}t}\right) \tag{4.13}$$

for $t \in [0, t_d]$ and by

$$C_B(t) = \frac{G}{V_B}(t - t_d) + C_B(t_d) \tag{4.14}$$

for $t \in (t_d, T]$.

For given values of G, D_M, D_r, t_d and T the temporal development of $C_B(t)$ will finally approach a stable state which is expressed by the fact that $C_B(t)$ is a T-periodical function of the time t. For this to be the case it is necessary and sufficient that

$$C_B(T) = C_B(0) \tag{4.15}$$

holds true.

For the following we assume $D_r = 0$ in order to facilitate the mathematical considerations.

Then we have, by (4.13) and (4.14)

$$C_B(T) = \frac{G}{V_B}(T - t_d) + C_B(0)e^{-\frac{D_M}{V_B}t_d} + \frac{G}{D_M}\left(1 - e^{-\frac{D_M}{V_B}t_d}\right)$$

and the requirement (4.15) turns out to be equivalent to

$$C_B(0) = \left[\frac{1}{D_M} + \frac{1}{V_B}\frac{T - t_d}{1 - e^{-\frac{D_M}{V_B}t_d}}\right]G. \tag{4.16}$$

For D_M we again choose the value 150 $[mg/min]$ and for V_B the value 13600 $[ml]$ (which will also be used later).
Then we obtain

$$C_B(0) = \left[0.0067 + 0.0000735\frac{T - t_d}{1 - e^{-0.011t_d}}\right]G.$$

If we assume, that dialysis is performed every second day, i.e., if we put $T = 48h = 2880min$, then for

$$\alpha(t_d) = 0.0067 + 0.0000735\frac{2880 - t_d}{1 - e^{-0.011t_d}}$$

the following values are obtained

t_d	$4h$	$5h$	$6h$	$7h$	$8h$
$\alpha(t_d)$	0.216	0.204	0.196	0.189	0.184

From (4.16) and (4.13) we deduce

$$C_B(t_d) = \beta(t_d) \cdot G \tag{4.17}$$

where

$$\beta(t_d) = \frac{1}{D_M} + \frac{1}{V_B}\frac{T - t_d}{e^{+\frac{D_M}{V_B}t_d} - 1} \tag{4.18}$$

For the above values of D_M, V_B and T we get

$$\beta(t_d) = 0.0067 + 0.0000735 \frac{2880 - t_d}{e^{+0.011 t_d} - 1}$$

which leads to the table

t_d	$4h$	$5h$	$6h$	$7h$	$8h$
$\beta(t_d)$	0.022	0.014	0.010	0.008	0.007

If we define the dialysis effect by

$$E(t_d) = \frac{C_B(t_d)}{C_B(0)} = \frac{\beta(t_d)}{\alpha(t_d)}, \tag{4.19}$$

then we obtain the following table

t_d	$4h$	$5h$	$6h$	$7h$	$8h$
$E(t_d)$	0.102	0.069	0.051	0.042	0.038

The numerical values in the above tables are not realistic for which reason the one-compartment model will be given up in the following. But they already indicate a tendency (to be confirmed in the refined model lateron), namely, that by extending the durance of dialysis to large times t_d (beyond $7h$) the dialysis effect is only slightly improved.

On using (4.16), (4.17), and (4.18) the dialysis effect (4.19) can also expressed in the form

$$E(t_d) = 1 - \left(\frac{1}{\frac{V_B}{D_M(T-t_d)} + \frac{1}{1-e^{-D_M/V_B t_d}}} \right). \tag{4.20}$$

This formula shows immediately that, for fixed values of V_B, T, and t_d, the value of $E(t_d)$ decreases, if D_M increases. This is to be expected in the same way as a decrease of $E(t_d)$, if t_d is increased where V_B, T, and D_M are kept fixed.

This, however, cannot be derived from (4.20) in an obvious way. But one can verify that

$$\frac{\partial E}{\partial t_d}(t_d, T, V_B, D_M) < 0.$$

4.1.3 The Temporal Development of the Toxin Concentration in the Blood with Ultrafiltration

If ultrafiltration is present, the total flux of the toxical substance out of the blood into the dialyzer is given by (4.8) with $Q_F \neq 0$ and we can no more assume that the total blood volume $V_B(t)$ is independent of the time t. We will, however, assume that $V_B(t)$ is a T-periodic function which is to be expected, if the whole process of hemodialysis becomes T-periodic. In order to obtain a simple model we assume that the body generates blood water at a constant rate of G_W $[ml/min]$ within the whole time interval $[0, T]$ in order to compensate for the loss of blood water during the period $[0, t_d]$ of dialysis. If $V_B(0) = V_B^0$, then the temporal development of $V_B(t)$ is given by

$$V_B(t) = \begin{cases} V_B^0 + (G_W - Q_F)t & \text{for } t \in [0, t_d), \\ V_B(t_d) + G_W(t - t_d) & \text{for } t \in [t_d, T]. \end{cases} \tag{4.21}$$

The assumption $V_B(T) = V_B(0) = V_B^0$ now leads to the relation

$$G_W = \frac{t_d}{T} Q_F \tag{4.22}$$

which connects the ultrafiltration rate with the blood water generation rate, if $V_B(t)$ is T-periodic. This relation which can also be written in the form $G_W T = Q_F t_d$ simply expresses the fact that the amount of blood water generated in the time interval $[0, T]$ is the same as the amount of blood water lost by ultrafiltration during $[0, t_d]$. Insertion of G_W from (4.22) into (4.21) yields

$$V_B(t) = \begin{cases} V_B^0 + \left(\frac{t_d}{T} - 1\right) t Q_F & \text{for } t \in [0, t_d), \\ V_B(t_d) + \frac{t_d}{T}(t - t_d) Q_F & \text{for } t \in [t_d, T] \end{cases} \tag{4.23}$$

and we obtain

$$\frac{d}{dt}(V_B(t) C_B(t)) = \left(\frac{t_d}{T} - 1\right) Q_F C_B(t) + \left(V_B^0 + \left(\frac{t_d}{T} - 1\right) t Q_F\right) \dot{C}_B(t) \text{ for } t \in (0, t_d)$$

and

$$\frac{d}{dt}(V_B(t)C_B(t)) = \frac{t_d}{T}Q_F C_B(t) + \left(V_B(t_d) + \frac{t_d}{T}(t - t_d)Q_F\right)\dot{C}_B(t) \text{ for } t \in (t_d, T).$$

We again assume that

$$C_D(t) = 0 \text{ for } t \in [0, t_d].$$

Then (4.9) can be rewritten in the form

$$\left(V_B^0 + \left(\frac{t_d}{T} - 1\right)tQ_F\right)\dot{C}_B(t) = \left[-D_M\left(1 - \frac{Q_F}{Q_B}\right) - \frac{t_d}{T}Q_F - D_r\right]C_B(t) + G \tag{4.24}$$

for $t \in (0, t_d)$ and

$$\left(V_B(t_d) + \frac{t_d}{T}(t - t_d)Q_F\right)\dot{C}_B(t) = -\left(D_r + \frac{t_d}{T}Q_F\right)C_B(t) + G \tag{4.25}$$

for $t \in (t_d, T)$.

With

$$\alpha = D_M\left(1 - \frac{Q_F}{Q_B}\right) + \frac{t_d}{T}Q_F + D_r \tag{4.26}$$

and

$$\beta = D_r + \frac{t_d}{T}Q_F \tag{4.27}$$

the general solution of (4.24) and (4.25) is given by

$$C_B(t) = C_B(0)\left(\frac{V_B(t)}{V_B^0}\right)^{\frac{\alpha T}{Q_F(T-t_d)}} + \frac{G}{\alpha}\left(1 - \left(\frac{V_B(t)}{V_B^0}\right)^{\frac{\alpha T}{Q_F(T-t_d)}}\right) \tag{4.28}$$

for $t \in [0, t_d]$ and by

$$C_B(t) = C_B(t_d)\left(\frac{V_B(t)}{V_B(t_d)}\right)^{-\frac{\beta T}{Q_F t_d}} + \frac{G}{\beta}\left(1 - \left(\frac{V_B(t)}{V_B(t_d)}\right)^{-\frac{\beta T}{Q_F t_d}}\right) \tag{4.29}$$

for $t \in [t_d, T]$, respectively.

From the requirement $C_B(T) = C_B(0)$ we therefore deduce

$$C_B(t_d)\left(\frac{V_B^0}{V_B(t_d)}\right)^{-\frac{\beta T}{Q_F t_d}} + \frac{G}{\beta}\left(1 - \left(\frac{V_B^0}{V_B(t_d)}\right)^{-\frac{\beta T}{Q_F t_d}}\right) = C_B(0)$$

where

$$C_B(t_d) = C_B(0)\left(\frac{V_B(t_d)}{V_B^0}\right)^{\frac{\alpha T}{Q_F(T-t_d)}} + \frac{G}{\alpha}\left(1 - \left(\frac{V_B(t_d)}{V_B^0}\right)^{\frac{\alpha T}{Q_F(T-t_d)}}\right)$$

whence

$$C_B(0) = \frac{P}{Q} \cdot G \tag{4.30}$$

with

$$P = \frac{1}{\beta}\left(1 - \left(\frac{V_B^0}{V_B(t_d)}\right)^{-\frac{\beta T}{Q_F t_d}}\right) + \frac{1}{\alpha}\left(1 - \left(\frac{V_B(t_d)}{V_B^0}\right)^{\frac{\alpha T}{Q_F(T-t_d)}}\right)\left(\frac{V_B^0}{V_B(t_d)}\right)^{-\frac{\beta T}{Q_F t_d}} \tag{4.31}$$

$$Q = 1 - \left(\frac{V_B(t_d)}{V_B^0}\right)^{\left(\frac{\alpha}{T-t_d} + \frac{\beta}{t_d}\right) T Q_F^{-1}} \tag{4.32}$$

We demonstrate (4.30) by a numerical example which shows that the values obtained for $C_B(0)$ and $C_B(t_d)$ for urea are as unrealistic as in the case without ultrafiltration. We choose $D_M = 150$ [mg/min], $Q_F = 10$ [ml/min], $Q_B = 200$ [ml/min] (as in Section 4.1.1), and $V_B^0 = 13600$ [ml] (as in Section 4.1.2). We further choose $T = 2880$ and $t_d = 360$ [min]. Then we obtain from (4.23)

$$\begin{aligned} V_B(t_d) &= 13600 + \left(\frac{360}{2880} - 1\right) \cdot 360 \cdot 10 \\ &= 13600 - 3150 = 10450 \text{ } [ml]. \end{aligned}$$

On choosing $D_r = 0$ we get from (4.26), (4.27)

$$\alpha = 150\left(1 - \frac{10}{200}\right) + \frac{360}{2880} \cdot 10 = 142.5 + 1.25 = 143.75,$$

$$\beta = \frac{360}{2880} \cdot 10 = 1.25$$

which leads to

$$\frac{\alpha T}{Q_F(T - t_d)} = 16.43$$

and

$$\frac{\beta T}{Q_F t_d} = 1$$

and

$$\left(\frac{V_B(t_d)}{V_B^0}\right)^{\frac{\alpha T}{Q_F(T-t_d)}} = 0.0132185$$

and

$$\left(\frac{V_B^0}{V_B(t_d)}\right)^{-\frac{\beta T}{Q_F t_d}} = 0.7683824.$$

Therefore

$$\begin{aligned}
P &= \frac{1}{1.25}(1 - 0.7683824) + \frac{1}{143.45}(1 - 0.0132185) \cdot 0.7683824 \\
&= 0.1852941 + 0.0052856 \\
&= 0.1905797 \\
Q &= 1 - 0.7683824^{17.43} \\
&= 0.9898661
\end{aligned}$$

hence

$$C_B(0) = 0.19253079 \cdot G.$$

From (4.28) we then obtain

$$\begin{aligned}
C_B(t_d) &= 0.19253079 \cdot 0.0132185 \cdot G + \frac{G}{143.75}(1 - 0.0132185) \\
&= 0.0094096 \cdot G
\end{aligned}$$

and by (4.29) we indeed get

$$\begin{aligned}
C_B(T) &= [0.0094096 \cdot 0.7683824 + \frac{1}{1.25}(1 - 0.7683824)]\, G \\
&= 0.1925242 \cdot G.
\end{aligned}$$

Without ultrafiltration we have obtained

$$C_B(T) = C_B(0) = 0.196 \cdot G$$

which again shows that ultrafiltration can be neglected.

4.2 A Two-Compartment Model

4.2.1 Derivation of the Model Equations

We already mentioned in Section 4.1 that the numerical values for the dialysis effect given by (4.19) in the one-compartment model are unrealistic. But also the linear increase of the toxin concentration during the dialysis-free interval $[t_d, T]$ given by (4.14) cannot be confirmed experimentally. Instead a steep increase of the concentration immediately after turning off the dialyzer can be observed which then goes over into a linear increase.

The steep increase right after the dialysis can be explained by the fact that the toxin (urea or creatinine) in the cellular part of the body has a concentration that is different from its concentration in the blood and that its value at the end of the dialysis is higher than the one in the blood. Consequently, diffusion through the cell membranes takes place from the cellular part into the blood which leads to the initially steep increase of the toxin concentration there.

If one assumes that the toxin after its generation gets immediately into the blood (as with urea, for instance, which is mainly created in the liver), then the toxin concentration in the blood after the dialysis will not only increase by virtue of the diffusion out of the cellular part of the body but also by the generation of toxin in the blood and will exceed the toxin concentration in the cellular part some time later so that at the beginning of the next dialysis a smaller value will appear there. This process will be given a more precise quantitative description later.

Uremic toxins are essentially contained in the body liquid which can be subdivided into the intercellular, the interstitial and the intravasal part. The latter two (at least for urea and creatinine) can be considered as one compartment, since they are closely connected by capillars. We call it the extracellular part and denote it by the subscript E. Consequently we subdivide the whole body liquid volume into the cellular part (denoted by the subscript C) and the extracellular part which are separated by the cell membranes. Through these diffusion takes place in both directions.

If one denotes the toxin concentration in the cellular and extracellular part by $C_C = C_C(t)$ and $C_E = C_E(t)$, respectively, and the clearance of the cell membranes (measured by $[ml/min]$) by D_C, then, by Fick's law, the flux of the toxical substance from the cellular into the extracellular part and vice versa is given by

$$F_C(t) = D_C(C_C(t) - C_E(t)). \tag{4.33}$$

The flux of the toxical substance from the extracellular part into the dialyzer across the membrane of the dialyzer is based on the formula (see Section 4.1.2)

$$F_E(t) = -D(t)C_E(t) \tag{4.34}$$

where

$$D(t) = \begin{cases} D_M(t) + D_r & \text{for} \quad 0 \leq t < t_d, \\ D_r & \text{for} \quad t_d \leq t < T. \end{cases} \tag{4.35}$$

By $D_M(t)$ and D_r we denote the clearance of the artificial and the natural kidneys, respectively (measured by $[ml/min]$). The period of dialysis is given by the time interval $[0, t_d]$ and T is the time after which dialysis is repeated. Let us denote by V_C and V_E the volume of the cellular and the extracellular liquid of the body (measured by $[ml]$), respectively. Then the temporal changes of the toxin concentration $C_C(t)$ and $C_E(t)$ in the cellular and extracellular part of the body, respectively, are governed by the two differential equations

$$V_C\dot{C}_C(t) = D_C(C_E(t) - C_C(t)), \tag{4.36}$$

$$V_E\dot{C}_E(t) = D_C(C_C(t) - C_E(t)) - D(t)C_E(t) + G \tag{4.37}$$

where G denotes the generation rate of the toxin (measured by $[mg/min]$) and is assumed to be constant and to be generated in the extracellular part of the body.

The right hand side of (4.37) can also be written in the form $F_C(t)+F_E(t)+G$ with $F_C(t)$ and $F_E(t)$ given by (4.33) and (4.34), respectively.

The equations (4.36), (4.37) have also been presented in [5] together with their solutions $C_C(t)$ and $C_E(t)$ on $[0, t_d]$ in the case of a constant clearance $D_M(t) = D_M,\ t \in [0, t_d]$, of the dialyzer membrane when $C_C(0)$ and $C_E(0)$ are prescribed.

The existence of T-periodic solutions in the case of a time-dependent clearance can be shown (see [2]).

In Section 4.3.1 we will describe a general method to calculate such solutions. In Section 4.3 it will be shown how they can be computed numerically by discretizing the differential equations (4.36), (4.37). For this purpose a realistic value of the unknown clearance D_C of the cell membranes has to be found. This will happen in Section 4.2.2.

4.2.2 Determination of the Clearance of the Cell Membranes for Urea

In the differential equations (4.36), (4.37) the quantities V_C, V_E, D_M, D_r, and G are accessible to measurement or experimental determination. The times t_d and T can also be measured. This is, however, not possible for the initial value $C_C(0)$. One can prove (see [2]) that, for every choice of the parameters $V_C, V_E, D_C, D_M, D_r, G, T$, and t_d, the system (4.36), (4.37) has exactly one pair $(C_C(t), C_E(t))$ of solutions with

$$C_C(t) > 0, \qquad C_E(t) > 0, \qquad \text{for all} \quad t \in [0, T]$$

and

$$C_C(T) = C_C(0), \qquad C_E(T) = C_E(0). \tag{4.38}$$

We will use this fact in order to determine the unknown clearance D_C for urea within realistic limits. For this purpose we assume that no rest clearance of the kidneys is present, i.e. $D_r = 0$. We at first consider the system (4.36), (4.37) in the time interval $[t_d, T]$. It is easy to show that the difference function

$$y(t) = C_E(t) - C_C(t)$$

satisfies the differential equation

$$\dot{y}(t) = -\eta y(t) + \gamma \tag{4.39}$$

where

$$\eta = D_C \left(\frac{1}{V_E} + \frac{1}{V_C} \right) \quad \text{and} \quad \gamma = \frac{G}{V_E}.$$

For a given initial value $y(t_d)$ the solution of (4.39) on $[t_d, T]$ reads

$$y(t) = y(t_d)e^{-\eta(t-t_d)} + \frac{\gamma}{\eta}\left(1 + e^{-\eta(t-t_d)}\right).$$

Realistic values for V_C and V_E are $V_C = 13600$ $[ml]$ and $V_E = 27200$ $[ml]$, respectively. We choose (as before) $T = 2880$ and $t_d = 360$ $[min]$.
By [5], p. 60 it is generally held that D_C, for urea, ranges between 700 and 900 $[ml/min]$. If we put $D_C = 700$, then we obtain $\eta \approx 0.077$ and $e^{-\eta(T-t_d)} \approx e^{-194.6} \approx 3.10^{-85} \approx 0$. This implies $y(T) = \frac{\gamma}{\eta}$ and, in connection with the condition (4.38) of T-periodicity, the relation

$$C_E(0) - C_C(0) \approx \frac{G}{D_C}\frac{V_C}{V_C + V_E}. \tag{4.40}$$

This can also be considered to be valid for other values of T and t_d whenever t_d is relatively small with respect to T (which is practically always the case).

The same holds also true for smaller values of D_C. For instance, for $D_C = 300$ $[ml/min]$ we obtain $\eta \approx 0.033$ and

$$e^{-\eta(T-t_d)} \approx e^{-83.4} \approx 6.02 \cdot 10^{-37} \approx 0$$

which also leads to (4.40).

In order to determine D_C one chooses a realistic value for $C_E(0)$, say $C_E(0) = 1.5$ $[mg/ml]$. Then one tentatively chooses a value for D_C and determines the rate G of toxin generation such that the given value of $C_E(0)$ and the corresponding value of $C_C(0)$ calculated from (4.40) turn out to be initial values of the T-periodic solution of (4.36), (4.37).

If we assume the clearance of the dialyzer to be given as the function

$$D_M(t) = C_0 e^{-0.001t} \qquad \text{for} \qquad t \in [0, t_d] \tag{4.41}$$

where $C_0 = 132$ $[ml/min]$, then we obtain the following table by this procedure:

D_C	400	450	500	550	600	$[ml/min]$
G	12.87	12.97	13.05	13.12	13.18	$[mg/min]$
$C_C(0)$	1.4786	1.4808	1.4826	1.4841	1.4854	$[mg/ml]$

Therefore, if we choose D_C within the range between 400 and 600 $[ml/min]$, then, for $C_E(0) = 1.5$ $[mg/min]$ the corresponding initial value $C_C(0)$ of the T-periodic solution of (4.36), (4.37), for G chosen such that (4.40) holds, turns out to be approximately equal to 1.48 $[mg/min]$. This does not determine the value of D_C very precisely.

However, if conversely G and D_C are prescribed, then the initial value $C_E(0)$ of the corresponding T-period solution of (4.36), (4.37) for a fixed choice of G does not depend very strongly on D_C. So for $G = 10$ $[mg/min]$ and $D_C = 500 \pm 100$ $[ml/min]$ we get

$$C_E(0) = 1.15 \begin{Bmatrix} +0.016 \\ -0.012 \end{Bmatrix} \; [mg/ml]$$

as being shown by the following table:

D_C	400	450	500	550	600	$[ml/min]$
$C_E(0)$	1.1654	1.1564	1.1491	1.1431	1.1382	$[mg/ml]$

In the following we will therefore uniformly choose $D_C = 500$ $[ml/min]$.

The initial value $C_E(0)$ of the C_E-part of a T-periodic solution of (4.36), (4.37) also not changes strongly, if not only D_C (within the range between 400 and 600 $[ml/min]$) but also G slightly changes.

We demonstrate this by the following table:

D_C	400	450	500	550	600	$[ml/min]$
G	9.90	9.94	10.00	10.03	10.06	$[mg/min]$
$C_E(0)$	1.154	1.149	1.149	1.147	1.145	$[mg/ml]$

Similar results are obtained for the case of a time-independent clearance of the dialyzer. Instead of (4.41) we choose

$$D_M(t) = D_M = 132 \; [ml/min] \qquad \text{for} \qquad t \in [0, t_d].$$

Again we choose V_C, V_E, T, and t_d as above. Then, for $G = 10$ $[mg/min]$, we obtain, by the method of calculating T-periodic solutions of (4.36), (4.37) in Section 4.3.1, the following results

D_C	400	500	600	$[ml/min]$
$C_E(0)$	1.0484	1.0341	1.0213	$[mg/ml]$
$C_C(0)$	1.0318	1.0207	1.0102	$[mg/ml]$

Since, by the results of Section 4.3.3, the T-periodic solutions of (4.36), (4.37) depend linearly on the toxin generation rate G, we obtain $C_E(0) = 1.5$ as initial value of the C_E-part of the T-periodic solution of (4.36), (4.37), if we replace $G = 10$ by

$$G = \frac{1.5}{1.0484} \cdot 10 = 14.31 \quad \text{for} \quad D_C = 400,$$
$$G = \frac{1.5}{1.0341} \cdot 10 = 14.51 \quad \text{for} \quad D_C = 500, \quad \text{and}$$
$$G = \frac{1.5}{1.0213} \cdot 10 = 14.69 \quad \text{for} \quad D_C = 600,$$

which leads to the following table

D_C	400	500	600	$[ml/min]$
G	14.31	14.51	14.69	$[mg/min]$
$C_C(0)$	1.476	1.481	1.484	$[mg/ml]$

4.3 Computation of Periodic Toxin Concentrations

4.3.1 The General Method

If we define a vector function $y(t) = (C_C(t), C_E(t))^T$, its time derivative $\dot{y}(t) = (\dot{C}_C(t), \dot{C}_E(t))^T$ and

$$A(t) = \begin{pmatrix} -D_C/V_C & D_C/V_C \\ D_C/V_E & -(D_C + D(t))/V_E \end{pmatrix}, \quad b = \begin{pmatrix} 0 \\ G/V_E \end{pmatrix}, \tag{4.42}$$

then the differential equations (4.36), (4.37) can be written in the form

$$\dot{y}(t) = A(t)y(t) + b. \tag{4.43}$$

For every initial vector $y^0 = (C_C^0, C_E^0)^T$ there is exactly one absolutely continuous vector function $y = y(t)$ with $y(0) = y^0$ which satisfies (4.43) for all $t \in (0, T) \setminus \{t_d\}$ and which is given by the formula of variation of the constants

$$y(t) = Y(t)\left\{y^0 + \int_0^t Y(s)^{-1} b \, ds\right\}, \quad t \in [0, T], \tag{4.44}$$

where $Y = Y(t)$ is a 2×2 matrix function (the so called fundamental matrix function) which satisfies the matrix differential equation

$$\dot{Y}(t) = A(t)\bar{Y}(t) \qquad \text{for all} \qquad t \in (0, T) \setminus \{t_d\} \tag{4.45}$$

and the initial condition

$$\dot{Y}(0) = E_2 = 2 \times 2 \ - \ \text{unit matrix.} \tag{4.46}$$

From the representation (4.44) it follows that $y = y(t)$ is T-periodic, if and only if

$$(E_2 - Y(T))\, y^0 = Y(T) \int_0^T Y(s)^{-1} b \, ds. \tag{4.47}$$

If one defines a vector function

$$\tilde{y}(t) = Y(t) \int_0^t Y(s)^{-1} b \, ds, \quad t \in [0, T],$$

then $\tilde{y} = \tilde{y}(t)$ is the unique absolutely continuous vector function which satisfies

$$\dot{\tilde{y}}(t) = A(t)\tilde{y}(t) + b \quad \text{for all} \quad t \in (0, T) \setminus \{t_d\} \tag{4.48}$$

and

$$\tilde{y}(0) = \Theta_2 = \text{ zero vector of } \mathbb{R}^2. \tag{4.49}$$

The calculation of the unique absolutely continuous T-periodic solution $y = y(t)$ of (4.43) for $t \in (0, T) \setminus \{t_d\}$ can be performed in four steps:

1. Determine the absolutely continuous solution $\tilde{y} = \tilde{y}(t)$ of (4.48), (4.49).
2. For $e^1 = (1, 0)^T$ and $e^2 = (0, 1)^T$ determine the absolutely continuous solution $y^i = y^i(t)$ of

$$\dot{y}^i(t) = A(t)y^i(t) \quad \text{for all} \quad t \in (0, T) \setminus \{t_d\} \tag{4.50}$$

   and

$$y^i(0) = e^i \quad \text{for} \quad i = 1, 2. \tag{4.51}$$

   Then the solution of (4.45), (4.46) is given by

$$Y(t) = \left(y^1(t), y^2(t)\right), \qquad t \in [0, T].$$

3. Solve the linear system

$$(E_2 - Y(T))\, y^0 = \tilde{y}(T) \tag{4.52}$$

   for $y^0 \in \mathbb{R}^2$.
4. Determine the unique absolutely continuous solution $y = y(t)$ of (4.43) for all $t \in (0, T) \setminus \{t_d\}$ with $y(0) = y^0$.

Based on physical reflections it is possible in step 2) to determine the matrix $Y(T)$ directly with the aid of the matrix $Y(t_d)$, if T is large compared with t_d (which practically is always the case) and if the clearance D_r of the natural kidneys is zero. For this purpose we define

$$\gamma_i = V_C y_1^i(t_d) + V_E y_2^i(t_d) \tag{4.53}$$

where $y^i(t) = \left(y_1^i(t), y_2^i(t)\right)^T$ for $i = 1, 2$.

Then γ_i is the total amount of the poison in C and E at the end of the process of dialysis with initial concentration

$$y_j^i = \begin{cases} 0 \text{ for } & i \neq j, \\ 1 \text{ for } & i = j, \end{cases} \qquad i, j = 1, 2.$$

The temporal development $y^1(t)$ and $y^2(t)$ for $t \in [t_d, T]$ means physically that with increasing t the initial, i.e., for $t = t_d$, amount γ_1 and γ_2, respectively, of the poison distribute uniformly in C and E so that we have

$$\lim_{T\to\infty} y_j^1(T) = \frac{\gamma_1}{V}$$

and

$$\lim_{T\to\infty} y_j^2(T) = \frac{\gamma_2}{V}$$

for $j = 1, 2$ where $V = V_C + V_E$. Therefore we can assume, for sufficiently large $T > t_d$, that

$$Y(T) \approx \frac{1}{V}\begin{pmatrix} \gamma_1 & \gamma_2 \\ \gamma_1 & \gamma_2 \end{pmatrix}. \tag{4.54}$$

Further we have

$$\gamma_1 < V_C \qquad \text{and} \qquad \gamma_2 < V_E, \tag{4.55}$$

since V_C and V_E is the amount of poison at the time $t = 0$ for $i = 1$ and $i = 2$, respectively, and is diminished in the time interval $[t_d, T]$ by the process of dialysis.

In a similar way $\tilde{y}(T) = \left(\tilde{C}_C(T), \tilde{C}_E(T)\right)^T$ can also be derived from $\tilde{y}(t_d) = \left(\tilde{C}_C(t_d), \tilde{C}_E(t_d)\right)^T$. At first we obtain from considerations in Section 4.2.2 that

$$\tilde{C}_E(T) - \tilde{C}_C(T) \approx \frac{G}{D_C}\frac{V_C}{V_E + V_C} \tag{4.56}$$

(see (4.40)).

Multiplying the first equation of (4.48) with V_C and the second with V_E and then adding both equation leads to

$$V_C\,\dot{\tilde{C}}_C(t) + V_E\,\dot{\tilde{C}}_E(t) = G \qquad \text{for} \qquad t \in (t_d, T).$$

This implies

$$V_C\,\tilde{C}_C(t) + V_E\,\tilde{C}_E(t) = G(t - t_d) + V_C\,\tilde{C}_C(t_d) + V_E\,\tilde{C}_E(t_d), \quad t \in [t_d, T],$$

hence

$$V_C\,\tilde{C}_C(T) + V_E\,\tilde{C}_E(T) = G(T - t_d) + V_C\,\tilde{C}_C(t_d) + V_E\,\tilde{C}_E(t_d). \tag{4.57}$$

If we put

$$M_d = V_C\,\tilde{C}_C(t_d) + V_E\,\tilde{C}_E(t_d),$$

then we obtain from (4.56) and (4.57)

$$\begin{aligned}
\tilde{C}_C(T) &\approx \frac{1}{V_C + V_E}\left[G(T - t_d) + M_d - \frac{G}{D_C}\frac{V_E V_C}{V_C + V_E}\right],\\
\tilde{C}_E(T) &\approx \frac{1}{V_C + V_E}\left[G(T - t_d) + M_d + \frac{G}{D_C}\frac{V_C^2}{V_C + V_E}\right].
\end{aligned} \tag{4.58}$$

The linear system (4.52) now reads explicitly

$$\begin{aligned}
\left(1 - \frac{\gamma_1}{V}\right)C_C^0 - \frac{\gamma_2}{V}C_E^0 &= \tilde{C}_C(T),\\
-\frac{\gamma_1}{V}C_C^0 + \left(1 - \frac{\gamma_2}{V}\right)C_E^0 &= \tilde{C}_E(T).
\end{aligned} \tag{4.59}$$

Subtracting the first equation from the second gives (as to be expected)

$$C_E^0 - C_C^0 = \tilde{C}_E(T) - \tilde{C}_C(T) \approx \frac{G}{D_C}\frac{V_C}{V_E + V_C},$$

hence

$$C_E^0 \approx C_C^0 + \frac{G}{D_C}\frac{V_C}{V_E + V_C}. \tag{4.60}$$

Insertion into the first equation of (4.59) and solving for C_C^0 leads to

$$C_C^0 \approx \frac{1}{V - \gamma_1 - \gamma_2}\left[\frac{-\gamma_2 G V_C}{D_C(V_E + V_C)} + V\tilde{C}_C(T)\right]. \tag{4.61}$$

4.3.2 The Case of Constant Clearance of the Dialyzer

We assume the clearance of the dialyzer to be constant, i.e.,

$$D_M(t) = D_M \qquad \text{for all} \qquad t \in [0, t_d).$$

We further assume that $D_r = 0$, i.e., there is no rest clearance of the kidneys. As unique solutions of the differential equations (4.36), (4.37) ($\Leftrightarrow$ (4.43)) on $(0, t_d)$ and the initial conditions

$$C_C(0) = C_C^0 \qquad \text{and} \qquad C_E(0) = C_E^0 \tag{4.62}$$

we then obtain

$$\begin{aligned} C_C(t) =& C_C^0 e^{-\delta t} + \frac{C_1\delta}{\delta + \lambda_1}\left(e^{\lambda_1 t} - e^{-\delta t}\right) \\ &+ \frac{C_2\delta}{\delta + \lambda_2}\left(e^{\lambda_2 t} - e^{-\delta t}\right) - \frac{\gamma}{\alpha+\beta}\left(1 - e^{-\delta t}\right), \\ C_E(t) =& C_1 e^{\lambda_1 t} + C_2 e^{\lambda_2 t} - \frac{\gamma}{\alpha+\beta}, \qquad t \in [0, t_d], \end{aligned} \tag{4.63}$$

where

$$\begin{aligned} \alpha &= -\frac{D_M + D_C}{V_E}, \quad \beta = \frac{D_C}{V_E}, \quad \gamma = \frac{G}{V_E}, \quad \delta = \frac{D_C}{V_C}, \\ \lambda_{1,2} &= -\frac{A}{2} \pm \sqrt{\frac{A^2}{4} - B}, \\ A &= \delta - \alpha, \qquad B = -\delta(\alpha+\beta), \\ C_1 &= \frac{1}{\lambda_2 - \lambda_1}\left[(\lambda_2 - \alpha)C_E^0 - \beta C_C^0 + \frac{\gamma(\lambda_2 - \lambda_1)}{\alpha+\beta} - \gamma\left(1 - \frac{\lambda_1}{\alpha+\beta}\right)\right], \\ C_2 &= \frac{1}{\lambda_2 - \lambda_1}\left[(\alpha - \lambda_1)C_E^0 + \beta C_C^0 + \gamma\left(1 - \frac{\lambda_1}{\alpha+\beta}\right)\right]. \end{aligned}$$

This implies in particular for the solution $\tilde{y}(t) = (\tilde{C}_C(t), \tilde{C}_E(t))^T$ of (4.48) and (4.49)

$$\begin{aligned} \tilde{C}_C(t) &= \frac{\tilde{C}_1\delta}{\delta+\lambda_1}\left(e^{\lambda_1 t} - e^{-\delta t}\right) + \frac{\tilde{C}_2\delta}{\delta+\lambda_2}\left(e^{\lambda_2 t} - e^{-\delta t}\right) - \frac{\gamma}{\alpha+\beta}\left(1 - e^{-\delta t}\right), \\ \tilde{C}_E(t) &= \tilde{C}_1 e^{\lambda_1 t} + \tilde{C}_2 e^{\lambda_2 t} - \frac{\gamma}{\alpha+\beta}, \qquad t \in [0, t_d], \end{aligned}$$

where

$$\tilde{C}_1 = \frac{\gamma(\lambda_2 - \alpha - \beta)}{(\lambda_2 - \lambda_1)(\alpha + \beta)},$$

and

$$\tilde{C}_2 = \frac{\gamma(\alpha + \beta + \lambda_1)}{(\lambda_2 - \lambda_1)(\alpha + \beta)}.$$

Herewith one can calculate

$$M_d = V_C \tilde{C}_C(t_d) + V_E \tilde{C}_E(t_d)$$

and $\tilde{C}_C(T), \tilde{C}_E(T)$ according to (4.58). For the calculation of C_C^0 and C_E^0 according to (4.60) and (4.61), respectively, we need the quantities γ_1 and γ_2 given by (4.53).

This requires the calculation of $y_1^1(t_d) = C_C^1(t_d)$, $y_2^1(t_d) = C_E^1(t_d)$ and $y_1^2(t_d) = C_C^2(t_d)$, $y_2^2(t_d) = C_E^2(t_d)$ where $C_C^i = C_C^i(t)$, $C_E^i = C_E^i(t)$ for $i = 1, 2$ are the solutions of (4.36), (4.37) on $(0, t_d)$ with $G = 0$ under the initial conditions

$$C_C^1(0) = 1, \; C_E^1(0) = 0 \quad \text{and} \quad C_C^2(0) = 0, \; C_2^2(0) = 1.$$

These can be obtained from (4.63).

4.3.3 Discretization of the Model Equations

If the clearance of the dialyzer is time dependent, then the model equations (4.36) and (4.37) cannot be solved explicitly on the interval $[0, t_d]$ and have to be solved numerically. The simplest way of doing this is to discretize the model equations by replacing the time derivatives $\dot{C}_C(t)$ and $\dot{C}_E(t)$ on the left-hand sides of (4.36) and (4.37) by difference quotients. For this purpose we choose a time stepsize $\Delta t > 0$ such that

$$t_d = K \cdot \Delta t \qquad \text{and} \qquad T = N \cdot \Delta t \tag{4.64}$$

for $K, N \in \mathbb{N}$ with $2 \leq K < N$.

The discretized model equations then read

$$\begin{aligned} V_C(C_C(t + \Delta t) - C_C(t)) &= -D_C \Delta t (C_C(t) - C_E(t)), \\ V_E(C_E(t + \Delta t) - C_E(t)) &= D_C \Delta t (C_C(t) - C_E(t)) - D(t)\Delta(t) C_E(t) + G\Delta t \end{aligned} \tag{4.65}$$

for $t \in \{k \cdot \Delta t : k = 0, \ldots N - 1\}$ and $D(t)$ given by (4.35).

If we define

$$x(t) = \begin{pmatrix} C_C(t) \\ C_E(t) \end{pmatrix} \qquad b = \begin{pmatrix} 0 \\ G\Delta t/V_E \end{pmatrix} \qquad B(t) = \begin{pmatrix} b_{11}\ b_{12} \\ b_{21}\ b_{22} \end{pmatrix} \tag{4.66}$$

where

$$\begin{aligned} &b_{11} = 1 - D_C\Delta t/V_C,\ b_{12} = D_C\Delta t/V_C, \\ &b_{21} = D_C\Delta t/V_E, \qquad b_{22} = 1 - (D(t) + D_C)\Delta t/V_E, \end{aligned} \tag{4.67}$$

then the difference equations (4.65) can also be written in the form

$$x(t + \Delta t) = B(t)x(t) + b. \tag{4.68}$$

We have shown in [4] that, under the assumption

$$\begin{aligned} &D_M(t) \leq D_M \qquad \text{for all} \qquad t \in [0, t_d) \\ &\text{and} \\ &\Delta t < min(V_C/D_C, V_E/(D_M + D_r + D_C)), \end{aligned} \tag{4.69}$$

the system (4.68) $\Leftrightarrow$ (4.65) has exactly one solution $x(t) = (C_C(t), C_E(t))^T$ such that

$$\begin{aligned} &C_C(t) > 0, \qquad C_E(t) > 0 \\ &\text{and} \\ &x(t + T) = x(t) \qquad \text{for all} \qquad t = k\Delta t, \qquad k = 0, 1, 2, \ldots \end{aligned} \tag{4.70}$$

For sufficiently small Δt in comparison to T this solution can be taken as a substitute of the corresponding T-periodic solution of (4.36), (4.37). In order to determine the T-periodic solution of (4.68) we at first define

$$t_k = k\Delta t \quad \text{and} \quad B_k = B(t_k) \quad \text{for} \quad k = 0, \ldots, N.$$

Then we obtain from (4.65)

$$\begin{aligned} x(t_1) &= B_0 x(t_0) + b_0, \qquad & b_0 &= b, \\ x(t_2) &= B_1 B_0 x(t_0) + b_1, \qquad & b_1 &= B_1 b_0 + b_0, \end{aligned}$$

and, in general,

$$x(t_k) = B_{k-1}B_{k-2}\dots B_0x(t_0) + b_{k-1}$$
$$b_{k-1} = B_{k-1}B_{k-2} + b_{k-2}, \tag{4.71}$$

and, finally,

$$x(t_N) = B_{N-1}B_{N-2}\dots B_0x(t_0) + b_{N-1},$$
$$b_{N-1} = B_{N-1}B_{N-2} + b_{N-2}.$$

For the T-periodic solution of the system (4.68) $x(t_0) \in \mathbb{R}^2$ has to be chosen such that it is a solution of the linear system

$$(I - Y_N)x(t_0) = b_{N-1} \tag{4.72}$$

where

$$I = \begin{pmatrix} 1 & 0 \\ 0 & 1 \end{pmatrix}, \qquad Y_N = B_{N-1}B_{N-2}\dots B_0. \tag{4.73}$$

If $x(t_0)$ has been calculated as solution of (4.72), then $x(t_1), \dots, x(t_N) = x(t_0)$ are obtained from (4.71) for $k = 1, \dots, N$.

The matrix Y_N in (4.72) given by (4.73) can be determined as follows:

1. One puts $e^1 = (1, 0)^T$ and computes
$$(y_1^1, y_2^1)^T = B_{N-1}B_{N-2}\dots B_0e^1. \tag{4.74}$$
2. One puts $e^2 = (0, 1)^T$ and computes
$$(y_1^2, y_2^2)^T = B_{N-1}B_{N-2}\dots B_0e^2. \tag{4.75}$$

Then

$$Y_N = \begin{pmatrix} y_1^1 & y_1^2 \\ y_2^1 & y_2^2 \end{pmatrix}. \tag{4.76}$$

As a consequence of the definition of $b_0 = b$ in (4.66) and b_{k-1} for $k = 2, \dots, N$ in (4.71) it follows that the solution $x(t_0)$ of (4.69) depends linearly on G. This can be used for the determination of the clearance D_C of the cell membranes as being described in Section 4.2.2. The procedure proposed there consisted of choosing a realistic value of $C_E(0)$ and a tentative value of D_C and then to determine G such that the initial values $C_E(t_0)$ and $C_C(t_0)$ of the

T-periodic solution of (4.65) corresponding to D_C and G coincide with $C_E(0)$ and $C_C(0)$, respectively, where $C_C(0)$ is such that (4.40) is satisfied.

In order to achieve this we choose D_C as above and put $G = 10$. Then we compute the corresponding T-periodic solution of (4.65) whose initial values we denote by $C_C(t_0, 10)$ and $C_D(t_0, 10)$. Then the value of G to be determined is given by

$$G = \frac{C_E(0)}{C_E(t_0, 10)} \cdot 10 \tag{4.77}$$

since, due to the linear dependence of $x(t_0)$ on G, we know that

$$C_E(0) = \frac{G}{10} C_E(t_0, 10).$$

Since also

$$C_C(0) = \frac{G}{10} C_C(t_0, 10),$$

the condition (4.40) is automatically satisfied.

Let us demonstrate this procedure by a numerical example. For V_C, V_E, T, t_d and $D_M(t)$ we make the same choice as in Section 4.2.2. As time stepsize we take $\Delta t = 1$ $[min]$.

Then for $G = 10$ $[mg/min]$ and $D_C = 400, 450, 500, 550, 600$ $[ml/min]$ we obtain the following table:

D_C	400	450	500	550	600
$C_C(t_0, 10)$	1.1487	1.1416	1.1358	1.1310	1.1271
$C_E(t_0, 10)$	1.1654	1.1564	1.1491	1.1431	1.1382

If we choose $C_E(0) = 1.5$ $[mg/ml]$, then, for G given by (4.77) we get the following table

D_C	400	450	500	550	600
G	12.87	12.97	13.05	13.12	13.18
$C_C(0)$	1.4786	1.4808	1.4826	1.4841	1.4854

This table has also been presented in Section 4.2.2 already.

4.3.4 Numerical Results for Urea

In this section we will present the (positive) T-periodic solutions of the system (4.36), (4.37) computed numerically by discretization (as in Section 4.3.3) and discuss their dependence on the parameters of the system.

As period T for the rhythm of dialysis we choose $T = 2880$ [*min*] (i.e., 48 hours). We assume the clearance of the dialyzer to be time dependent of the form (4.35) with $D_M(t)$ given by (4.41) where we choose, for C_0, the values 132, 160, and 180 [*ml*/*min*].

Further we put (as in Section 4.2.2) $V_E = 13600, V_C = 27200$ [*ml*]. For the clearance D_C of the cell membranes we take the value 500 [*ml*/*min*] which is justified by the considerations in Section 4.2.2. This value refers to urea whose generation rate G we assume to be 10 [*mg*/*min*]. For the beginning we assume that there is no rest clearance of the natural kidneys and put $D_r = 0$.

The discretization of the system (4.36), (4.37) was performed by the Runge-Kutta method (and not by Euler's polygon method as being described in Section 4.3.3) with stepsize $\Delta t = 1$ [*min*].

In the following we only present the values of $C_E(t)$ for these can also be measured. In Table 4.1 we present some values of $C_E(t)$ during the time interval of dialysis and in dependence of C_0 and t_d.

C_0	132	132	132	160	160	160	181	181	181
t_d	300	360	420	300	360	420	300	360	420
$C_E(0)$	1.284	1.149	1.052	1.144	1.032	0.950	1.069	0.969	0.896
$C_E(60)$	0.981	0.879	0.806	0.828	0.748	0.690	0.744	0.675	0.625
$C_E(120)$	0.851	0.764	0.702	0.700	0.633	0.585	0.617	0.651	0.520
$C_E(180)$	0.749	0.673	0.619	0.601	0.545	0.505	0.521	0.475	0.441
$C_E(240)$	0.665	0.600	0.553	0.523	0.475	0.441	0.446	0.408	0.380
$C_E(300)$	0.597	0.539	0.498	0.460	0.420	0.390	0.388	0.355	0.332
$C_E(360)$		0.489	0.453		0.375	0.349		0.314	0.294
$C_E(420)$			0.425			0.316			0.264
$\frac{C_E(t_d)}{C_E(0)}$	0.465	0.426	0.404	0.402	0.363	0.333	0.363	0.324	0.295

Table 4.1.

In the last row we have listed the values of the dialysis effect

$$E(C_0, t_d) = \frac{C_E(t_d)}{C_E(0)} \tag{4.78}$$

which renders to be monotonically decreasing with respect to C_0 (for t_d being fixed) and with respect to t_d (for C_0 being fixed). This has also been observed in Section 4.1.2 for the one-compartment model (the values there, however, are too small).

If C_0 is fixed, then the improvement of the dialysis effect is higher in passing from 5 to 6 hours than from 6 to 7 hours as period of the process of dialysis. This shows that one should rather increase the value of C_0 in order to achieve an improvement of the dialysis effect.

Table 4.2 shows the influence of the rest clearance D_r of the natural kidneys to the C_E-part of the T-periodic solution of (4.36), (4.37). We choose in particular $t_d = 360$ [min] and $C_0 = 132$ [ml/min].

D_r	1	2	3	4	5	6
$C_E(0)$	1.061	0.985	0.918	0.860	0.808	0.761
$C_E(60)$	0.812	0.753	0.702	0.658	0.618	0.582
$C_E(120)$	0.706	0.665	0.611	0.572	0.538	0.507
$C_E(180)$	0.622	0.578	0.539	0.505	0.475	0.448
$C_E(240)$	0.554	0.515	0.481	0.450	0.424	0.400
$C_E(300)$	0.499	0.464	0.433	0.406	0.382	0.361
$C_E(360)$	0.453	0.422	0.394	0.370	0.349	0.329
$\frac{C_E(t_d)}{C_E(0)}$	0.427	0.428	0.429	0.430	0.432	0.432

Table 4.2.

Then we obtain Table 4.2 which shows that in comparison with $D_r = 0$ the dialysis effect does not improve (it even deteriorates) and gets worse with increasing values of D_r. But the maximum values $C_E(0)$ of $C_E(t)$ become significantly smaller with growing values of D_r. This is understandable, since the natural kidneys are permanently active (i.e., 48 hours within a dialysis period of 48 hours) whereas the artificial kidney only acts for 6 hours.

If the rest clearance of thc natural kidneys is large enough, then dialysis can be renounced, i.e. we can put $D_M(t) = 0$ for $t \in [0, t_d)$ such that

$$D(t) = D_r \qquad \text{for all} \qquad t \in [0, T].$$

The model equations (4.36), (4.37) then possess unique T-periodic solutions which are constant, i.e., $C_C(t) = C_C, C_E(t) = C_E$ for all $t \in [0, T]$ where C_C and C_E are the unique solutions of the linear system

$$D_C(C_E - C_C) = 0,$$
$$D_C(C_C - C_E) - D_r C_E + G = 0$$

and are given by

$$C_C = C_E = \frac{G}{D_r}.$$

The presence of some rest clearance of the natural kidneys can also be used in order to enlarge the time T between subsequent periods of dialysis. The Table 4.3 gives some values of the C_E-part of the T-periodic solution of (4.36), (4.37) for $T = 4320$ $[min]$ (3 days), $t_d = 360$ $[min]$, and for several values of C_0 and D_r.

C_0	132	132	160	160	181	181
D_r	0	1.3	0	0.5	0	0.5
$C_E(0)$	1.748	1.502	1.572	1.492	1.478	1.409
$C_E(60)$	1.331	1.143	1.133	1.075	1.023	0.975
$C_E(120)$	1.150	0.989	0.954	0.905	0.845	0.805
$C_E(180)$	1.008	0.866	0.815	0.774	0.709	0.676
$C_E(240)$	0.891	0.766	0.705	0.669	0.603	0.576
$C_E(300)$	0.795	0.685	0.616	0.585	0.520	0.496
$C_E(360)$	0.716	0.617	0.544	0.517	0.454	0.433

Table 4.3.

If $C_E(0) = 1.5$ $[mg/min]$ can be tolerated as maximum value of $C_E(t)$, then for $C_0 = 132$ $[ml/min]$ a rest clearance $D_r = 1.3$ $[ml/min]$ would suffice whereas for $C_0 = 160$ $[ml/min]$ the value $D_r = 0.5$ $[ml/min]$ would be sufficient and for $C_0 = 181$ $[ml/min]$ the rest clearance D_r of the kidneys could be zero.

4.3.5 The Influence of the Urea Generation Rate

If we define (see Section 4.3.1)

$$\begin{aligned} A(t) &= \begin{pmatrix} -D_C/V_C & D_C/V_C \\ D_C/V_E & -\left(D_C + D(t)\right)/V_E \end{pmatrix}, \\ b &= \begin{pmatrix} 0 \\ G/V_E \end{pmatrix}, \quad x(t) = \begin{pmatrix} C_C(t) \\ C_E(t) \end{pmatrix}, \quad \text{and} \quad \dot{x}(t) = \begin{pmatrix} \dot{C}_C(t) \\ \dot{C}_e(t) \end{pmatrix}, \end{aligned} \tag{4.79}$$

then the system (4.36), (4.37) can also be written in the form

$$\dot{x}(t) = A(t)x(t) + b \tag{4.80}$$

and has, for every given $x(0) = x_0 = (C_C^0, C_E^0)^T$, exactly one (absolutely continuous) solution which can be represented by the formula of variation of the constants as

$$x(t) = Y(t)\left\{x_0 + \int_0^t Y(s)b\,ds\right\}, \quad t \in [0, T], \tag{4.81}$$

where $Y = Y(t)$ is the so called fundamental matrix function which satisfies

$$\dot{Y}(t) = A(t)Y(t) \qquad \text{for all} \qquad t \in (0, T) \setminus \{t_d\} \tag{4.82}$$

and

$$Y(0) = I_2 = 2 \times 2 - \text{unit matrix}.$$

If $x = x(t)$ is T-periodic, i.e., $x(T) = x_0$, then it follows that

$$x_0 = (I_2 - Y(T))^{-1} Y(T) \int_0^T Y(s)^{-1} b\,ds \tag{4.83}$$

(The existence of $(I_2 - Y(T))^{-1}$ is guaranteed by virtue of the investigations in [2]). By inserting x_0 from (4.83) into (4.81) one recognizes that every T-periodic solution $(C_C(t), C_E(t))^T$ of the system (4.36), (4.37) depends linearly on G.

The same holds true for the T-periodic solutions of the discretized model equations (4.65) which are taken as approximate solutions of (4.36), (4.37). This is shown in Section 4.3.2. The existence of T-periodic solutions of (4.65) is proved in [4] under the assumption (4.69).

If we denote the dependence of $C_E(t)$ on G by $C_E(t, G)$, then we have

$$C_E(t, G) = G \cdot C_E(t, 1) \tag{4.84}$$

which is equivalent to

$$\frac{C_E(t, G_1)}{C_E(t, G_2)} = \frac{G_1}{G_2}. \tag{4.85}$$

It therefore suffices, to compute the C_E-part of a T-periodic solution of (4.36), (4.37) for some value of G, say $G = 10$ (as in Section 4.3.4). The corresponding C_E-part for any value G can then be determined by virtue of formula (4.85).

4.3.6 Determination of the Urea Generation Rate and the Rest Clearance of the Kidneys

We assume that for some patient who is dialysed for 6 hours every two days a periodic development of $C_E(t)$ has established with the period of $T = 2880$ $[min]$. If the rest clearance D_r of his kidneys is known (for instance, $D_r = 0$), then the urea generation rate G can be determined experimentally with the aid of the relation (4.85) as follows: At the beginning of the dialysis period the value $C_E(0, G)$ is measured and the T-periodic solution $(C_C(t, \tilde{G}), C_C(t, \tilde{G}))^T$ of (4.36), (4.37) is calculated for some $\tilde{G}$, say $\tilde{G} = 10$. The unknown rate G is then given, by virtue of (4.85), as

$$G = 10\frac{C_E(0, G)}{C_E(0, 10)}. \tag{4.86}$$

If the rest clearance D_r of the kidneys is unknown, then one can proceed as follows: At the beginning of the period of dialysis $C_E(0) = CE_0$ is measured. Then, at least two hours after the end of the dialysis, urine of the patient is collected within some time interval $[t_a, t_e]$ (where, for instance $t_e - t_a = 1440$ $[min]$) and the amount H_g of urea contained in the urine is determined. From this the amount of urea excreted per minute within the interval $[t_a, t_e]$ is obtained as $H = H_g/(t_e - t_a)$.

If D_r and G are the quantities to be determined, then for the C_E-part of the corresponding T-periodic solution $(C_C(t, G, D_r), C_E(t, G, D_r))^T$ of (4.36), (4.37) we must require that

$$H = \frac{1}{t_e - t_a} \int_{t_a}^{t_e} D_r C_E(t, G, D_r)\, dt. \tag{4.87}$$

and

$$C_E(0, G, D_r) = CE_0.$$

For $G = 10$ $[mg/min]$ let

$$\begin{aligned} \overline{CE}(D_r) &= \frac{1}{t_e - t_a} \int_{t_a}^{t_e} C_E(t, 10, D_r)\, dt \\ &\approx \frac{1}{6} \left\{ C_E(t_a, 10, D_r) + 4C_E\left(\frac{t_a + t_e}{2}, 10, D_r\right) + C_E(t_e, 10, D_r) \right\}. \end{aligned} \tag{4.88}$$

From (4.84) we deduce

$$\frac{C_E(t, 10, D_r)}{C_E(0, 10, D_r)} = \frac{C_E(t, G, D_r)}{C_E(0, G, D_r)} = \frac{C_E(t, G, D_r)}{CE_0}$$

such that from (4.87) and (4.88) we obtain the relation

$$\frac{\overline{CE}(D_r)}{C_E(0, 10, D_r)} = \frac{1}{D_r} \cdot \frac{H}{CE_0}. \tag{4.89}$$

If we choose $C_0 = 132$ $[ml/min]$, $t_d = 360$ $[min]$, $t_a = 660$, and $t_e = 2100$ $[min]$, then for

$$g(D_r) = \frac{\overline{CE}(D_r)}{C_E(0, 10, D_r)}$$

we obtain the following table:

D_r	0	1	2	3	4	5
$g(D_r)$	0.6801	0.6843	0.6885	0.6927	0.6970	0.7012

From this table we derive the linear relation $g(D_r) = aD_r + b$ with $a = 0.0042, b = 0.6801$ which, in connection with (4.89), leads to the quadratic equation

$$aD_r^2 + bD_r - \frac{H}{CE_0} = 0$$

whose positive solution is given by

$$D_r = -\frac{b}{2a} + \sqrt{\frac{b^2}{4a^2} + \frac{H}{CE_0}}. \tag{4.90}$$

Having determined D_r from this formula we get G, in analogy to (4.86), from

$$G = 10 \cdot \frac{CE_0}{C_E(0, 10, D_r)}.$$

4.4 A Three-Compartment Model

4.4.1 Motivation and Derivation of the Model Equations

Headaches that occur at the beginning of the period after the process of dialysis with numerous patients give rise to the assumption that the cell membranes of the brain have a smaller clearance than the other cell membranes of the body. This implies that the concentration of the toxin in the brain during the final part of the process of dialysis and also shortly afterwards is considerably higher than in the extracellular part of the body in which the toxin concentration is reduced by the dialyzer. This surplus of toxin concentration in the brain may be the reason for the observed headache.

Therefore we will treat the brain as a separate part of the cellular part of the body. Then besides the diffusion between the latter and the extracellular part we have also to regard a diffusion between the brain and the extracellular part of the body. The two differential equations (4.36), (4.37)) therefore have to be replaced by the following three differential equations:

$$V_C\dot{C}_C(t) = D_C(C_E(t) - C_C(t)), \tag{4.91}$$

$$V_B\dot{C}_B(t) = D_B(C_E(t) - C_B(t)), \tag{4.92}$$

$$V_E\dot{C}_E(t) = D_C(C_C(t) - C_E(t)) + D_B(C_B(t) - C_E(t)) - D(t)C_E(t) + G. \tag{4.93}$$

In addition to the quantities that have been introduced already in Section 4.2.1 we have the following: V_B = volume of the brain (in ml), D_B = clearance of the brain cells (in ml/min), $C_B(t)$ = toxin concentration in the brain at the time t (in mg/ml) and $\dot{C}_B(t)$ = time derivative of $C_B(t)$ at the time t. We also assume $D_r = 0$.

The equation (4.91) and (4.92) describes the temporal change of C_C and C_B under the influence of diffusion between C and E and B and E, respectively. Equation (4.93) describes the temporal change of C_E under the influence of diffusion between C, B and E, under the influence of intermittent dialysis and caused by the generation of the toxin.

This three-compartment model reduces to the two-compartment model (of Section 4.2.1), if we require that $C_B(t) = C_C(t) = C(t)$ which must be the case, if the brain is no more separated from the cellular part of the body. From (4.91) and (4.92) we then necessarily derive the relation

$$\frac{D_C}{V_C} = \frac{D_B}{V_B} \tag{4.94}$$

which implies

$$D_B + D_C = \frac{V_C + V_B}{V_C} D_C \tag{4.95}$$

and further (by addition of (4.91) and (4.92))

$$(V_C + V_B)\dot{C}(t) = D(C_E(t) - C(t)) \tag{4.96}$$

with

$$D = D_B + D_C. \tag{4.97}$$

This is exactly the equation (4.36) with $V_C + V_B$ instead of V_C and D instead of D_C and $C(t)$ instead of $C_C(t)$, respectively. Equation (4.93) can be rewritten in the form

$$V_E\dot{C}_E(t) = D(C(t) - C_E(t)) - D(t)C_E(t) + G \tag{4.98}$$

which is exactly the equation (4.37) with D instead of D_C and $C(t)$ instead of $C_C(t)$.

By [3] there exist T-periodic solutions

$$C_C(t) > 0, \qquad C_B(t) > 0, \qquad C_E(t) > 0, \qquad \text{for} \qquad 0 \leq t \leq T$$

of the system (4.94) the calculation of which will be discussed in Section 4.4.3. Before, however, we have to determine realistic values for the clearance D_B of the brain cells. This will happen in Section 4.4.2.

4.4.2 Determination of the Clearance of the Cell Membranes of the Brain

As in Section 4.3.4 we again choose $T = 2880$ and $t_d = 360$ [min],

$$C_M(t) = 132e^{-0.001t}, \qquad t \in [0, t_d],$$

$G = 10$ [mg/min], $V_E = 13600$ [ml]. As volume of the brain we choose $V_B = 1375$ and for the rest of the cellular part of the body the volume $V_C = 25825$ [ml] so that on the whole we again obtain $V_C + V_B = 27200$ [ml].

A first hint to the order of size of D_B can be derived from the two-compartment model (4.96), (4.98). By the investigations of Section 4.2.2 it is reasonable to assume that $D = 500$ [ml/min] (which corresponds to $D_C = 500$ [ml/min]). Then (4.95) and (4.97) imply

$$D_C = \frac{V_C D}{V_C + V_B} = \frac{25825 \cdot 500}{27200} = 474.72$$

and further

$$D_B = D - D_C = 25.28 \ [ml/min].$$

In general, (4.94) implies

$$D_B = \frac{V_B}{V_C} D_C.$$

If $T - t_d$ is sufficiently large, then one can show the two relations

$$C_E(T) - C_C(T) \approx \frac{G V_C}{(V_C + V_B + V_E) D_C} \tag{4.99}$$

and

$$C_E(T) - C_B(T) \approx \frac{G V_B}{(V_C + V_B + V_E) D_B}. \tag{4.100}$$

By (4.94), these two equations reduce to one equation in the case of a two-compartment model (where $C_C(T) = C_B(T) = C(T)$ and $D_C = D_B = D$), namely,

$$C_E(T) - C(T) \approx \frac{G}{D} \frac{V_C}{V_C + V_B + V_E}, \tag{4.101}$$

which, for T-periodic solutions (i.e. in the case $C_E(T) = C_E(0)$ and $C(T) = C(0)$) is exactly the equation (4.40) with D instead of D_C and $V_C + V_B$ instead of V_C.

Based on the reasonable assumption that

$$C_B(T) \leq C_C(T) \leq C_E(T) \tag{4.102}$$

one then derives from (4.99) and (4.100) that

$$D_B \leq \frac{V_B}{V_C} D_C = 25.28 \tag{4.103}$$

where equality occurs in the case of the two-compartment model as been shown above.

4.4.3 Computation of Periodic Urea Concentration Curves

If we define

$$A(t) = \begin{pmatrix} -D_C/V_C & 0 & 0 \\ 0 & -D_B/V_B & 0 \\ D_C/V_E & D_B/V_E & -(D_C + D_B + D(t))/V_E \end{pmatrix},$$

$$b = \begin{pmatrix} 0 \\ 0 \\ G/V_E \end{pmatrix}, \qquad y = \begin{pmatrix} C_C(t) \\ C_B(t) \\ C_E(t) \end{pmatrix}, \qquad \dot{y}(t) = \begin{pmatrix} \dot{C}_C(t) \\ \dot{C}_B(t) \\ \dot{C}_E(t) \end{pmatrix},$$

then the system (4.91), (4.92), (4.93) can be rewritten in the form

$$\dot{y}(t) = A(t)y(t) + b. \tag{4.104}$$

For each initial vector $y_0 \in \mathbb{R}^3$ it is known (see, for instance, [3]) that there is exactly one absolutely continuous solution $y = y(t) \in \mathbb{R}^3$ for $t \in [0, T]$ such that $y(0) = y_0$ which satisfies (4.104) for all $t \in (0, T) \setminus \{t_d\}$. This is given by the formula of variation of the constants

$$y(t) = Y(t)\{y_0 + \int_0^t Y(s)^{-1} b \, ds\} \tag{4.105}$$

for $t \in [0, T]$ where $Y = Y(t)$ is the so called fundamental matrix function on $[0, T]$ with

$$\dot{Y}(t) = A(t)Y(t) \qquad \text{for} \qquad t \in (0, T) \setminus \{t_d\} \tag{4.106}$$

and

$$Y(0) = E_3 = 3 \times 3 - \text{unit matrix}. \tag{4.107}$$

For the T-periodic solution $y = y(t)$ of (4.104) (which exists by [3] and is unique) then necessarily

$$y_0 = Y(t)\{y_0 + \int_0^T Y(t)^{-1} b dt\}$$

holds true which leads to the linear system

$$(E_3 - Y(T))y_0 = \tilde{y}(T) \tag{4.108}$$

for $y_0 \in \mathbb{R}^3$ where

$$\tilde{y}(t) = Y(t) \int_0^t Y(s)^{-1} b \, ds \qquad \text{for} \qquad t \in [0, T]. \tag{4.109}$$

The computation of the T-periodic solution $y = y(t), t \in [0, T]$, of (4.104) now happens in 4 steps.

Step 1: At first $\tilde{y} = \tilde{y}(t)$ given by (4.109) is determined as the unique absolutely continuous solution of the initial value problem

$$\begin{aligned} \dot{\tilde{y}}(t) &= A(t)\tilde{y}(t) + b, \quad t \in (0, T), \\ \tilde{y}(0) &= \Theta_3 = \text{zero vector in } \mathbb{R}^3 \end{aligned} \tag{4.110}$$

Step 2: Next $Y = Y(t) = (y^1(t), y^2(t), y^3(t)), y^i(t) \in \mathbb{R}^3$ for $i = 1, 2, 3$ and $t \in [0, T]$ which satisfies (4.106) and (4.107) is determined by solving the following three initial value problems:

$$\begin{aligned} \dot{y}^i(t) &= A(t)y^i(t), \qquad & t \in (0, T), \\ y^i &= e^i, & i = 1, 2, 3, \end{aligned} \tag{4.111}$$

where e^i denotes the i-th coordinate unit vector in $\mathbb{R}^3$ (i.e., $e^i_j = \delta_{ij} =$ Kronecker's symbol for $i, j = 1, 2, 3$).

Step 3: Third the unique solution $y_0 \in \mathbb{R}^3$ of (4.108) is calculated.

Step 4: Finally, the (componentwise) positive, absolutely continuous, T-periodic solution of (4.104) (if $y_0 \in \overset{\circ}{\mathbb{R}}{}^3_+$) is determined by solving the initial value problem

$$\begin{aligned} \dot{y}(t) &= A(t)y(t) + b, \\ y(0) &= y_0 \end{aligned} \tag{4.112}$$

with $y_0 \in \mathbb{R}^3$ taken from step 3.

The initial value problems in steps 1, 2, and 4 can be solved numerically as in Section 4.3.3 by discretization and Euler's polygon method. In this way the numerical results presented in Section 4.4.4 were obtained (by choosing as step length $\Delta t = 1$ [min]).

For T and D_B being sufficiently large the computation of $\tilde{y}(T)$ in step 1 and of $Y(T)$ in step 2 can be performed more economically than by solving (4.110), (4.111) and (4.112), respectively, by means of Euler's polygon method.

We begin with the determination of $\tilde{y}(T)$ in step 1. At first we determine, by Euler's polygon method, $\tilde{y}(t_d) = (\tilde{y}_1(t_d), \tilde{y}_2(t_d), \tilde{y}(t_d))^T$ and calculate

$$\tilde{\gamma} = V_C \tilde{y}_1(t_d) + V_B \tilde{y}_2(t_d) + V_E \tilde{y}_3(t_d). \tag{4.113}$$

Similar to (4.58) it is possible to derive, for $\tilde{y}(T) = (\tilde{C}_C(T), \tilde{C}_B(T), \tilde{C}_E(T))^T$ the relations

$$\tilde{C}_C(T) \approx \tilde{C}_E(T) - \frac{GV_C}{(V_C + V_B + V_E)D_C}, \tag{4.114}$$

$$\tilde{C}_B(T) \approx \tilde{C}_E(T) - \frac{GV_B}{(V_C + V_B + V_E)D_B}, \tag{4.115}$$

$$\tilde{C}_E(T) \approx \frac{1}{V_C + V_B + V_E}\left[G(T - t_d) + \tilde{\gamma} + \left(\frac{V_C^2}{D_C} + \frac{V_B^2}{D_B}\right)\frac{G}{V_C + V_B + V_E}\right]. \tag{4.116}$$

Next we consider the determination of $Y(T)$. In analogy to (4.113) we define

$$\gamma_i = V_C y_1^i(t_d) + V_B y_2^i(t_d) + V_E y_3^i(t_d) \tag{4.117}$$

for $i = 1, 2, 3$. Then γ_i is the total amount of urea at the end of dialysis corresponding to the initial urea concentration $y_j^i(0) = \delta_{ij}$ for $j = 1, 2, 3$.

The relations (4.117) can also be written in the form

$$\begin{pmatrix}\gamma_1\\ \gamma_2\\ \gamma_3\end{pmatrix} = \overbrace{\begin{pmatrix} y_1^1(t_d) & y_2^1(t_d) & y_3^1(t_d)\\ y_1^2(t_d) & y_2^2(t_d) & y_3^2(t_d)\\ y_1^3(t_d) & y_2^3(t_d) & y_3^3(t_d)\end{pmatrix}}^{Y(t_d)^T} \begin{pmatrix}V_C\\ V_B\\ V_E\end{pmatrix}$$

Since the computation of $Y(t)$ for $t \in [t_d, T]$ is done under the assumption that $G = 0$ and

$$D(t) = 0 \qquad \text{for} \qquad t \in [t_d, T]$$

(see (4.35) and observe that we assume $D_r = 0$), the total amount γ_i of urea for $i = 1, 2, 3$ at $t = t_d$ will be uniformly distributed over the extracellular part of the body, the brain, and the rest of the cellular part as t tends to infinity. So, for $t \to \infty$ in all three parts there will be the same urea concentration γ_i/V with $V = V_C + V_B + V_E$. This means that

$$\lim_{t\to\infty} Y(t) = \frac{1}{V}\begin{pmatrix}\gamma_1 & \gamma_2 & \gamma_3\\ \gamma_1 & \gamma_2 & \gamma_3\\ \gamma_1 & \gamma_2 & \gamma_3\end{pmatrix}. \tag{4.118}$$

Further it is clear that

$$\gamma_1 < V_C, \qquad \gamma_2 < V_B, \qquad \gamma_3 < V_E, \tag{4.119}$$

V_C and V_B and V_E being the total amount of urea at $t = 0$ for $i = 1$ and $i = 2$ and $i = 3$, respectively which is diminished in the course of dialysis. From (4.118) and (4.119) it follows that

$$\lim_{t\to\infty} Y(t)^T \begin{pmatrix}V_C\\ V_B\\ V_E\end{pmatrix} = \begin{pmatrix}\gamma_1\\ \gamma_2\\ \gamma_3\end{pmatrix} < \begin{pmatrix}V_C\\ V_B\\ V_E\end{pmatrix}.$$

On using a so called Quotient-Theorem for the spectral radius $\rho(Y(T))$ of $Y(T)$ (see, for instance, [1]) it follows, for sufficiently large T, that $\rho(Y(T)) < 1$ which proves the unique solvability of (4.108), if T is sufficiently large.

If we assume T so large that

$$Y(T) \approx \lim_{t\to\infty} Y(t) = \frac{1}{V}\begin{pmatrix} \gamma_1 & \gamma_2 & \gamma_3 \\ \gamma_1 & \gamma_2 & \gamma_3 \\ \gamma_1 & \gamma_2 & \gamma_3 \end{pmatrix}, \tag{4.120}$$

then (4.108) can be explicitly solved for $y_0 = (C_C^0, C_B^0, C_E^0)^T$. In order to see this we rewrite (4.108) in the form

$$\begin{aligned} (V-\gamma_1)C_C^0 - \gamma_2 C_B^0 - \gamma_3 C_E^0 &= V\tilde{C}_C(T) \\ -\gamma_1 C_C^0 + (V-\gamma_2)C_B^0 - \gamma_3 C_E^0 &= V\tilde{C}_B(T) \\ -\gamma_1 C_C^0 - \gamma_2 C_B^0 + (V-\gamma_3)C_E^0 &= V\tilde{C}_E(T) \end{aligned} \tag{4.121}$$

On subtracting the third equation from the first and the second we obtain

$$\begin{aligned} V(C_C^0 - C_E^0) &= V(\tilde{C}_C(T) - \tilde{C}_E(T)) \\ V(C_B^0 - C_E^0) &= V(\tilde{C}_B(t) - \tilde{C}_E(T)) \end{aligned}$$

which leads to (see (4.114) and (4.115))

$$C_C^0 = C_E^0 - \frac{GV_C}{(V_C + V_B + V_E)D_C}, \tag{4.122}$$

$$C_B^0 = C_E^0 - \frac{GV_C}{(V_C + V_B + V_E)D_B}. \tag{4.123}$$

Insertion of

$$C_C^0 = C_E^0 + (\tilde{C}_C(T) - \tilde{C}_E(T)) \quad \text{and} \quad C_B^0 = C_E^0 + (\tilde{C}_B(T) - \tilde{C}_E(T))$$

into the first equation of (4.121) yields

$$\begin{aligned} &(V-\gamma_1-\gamma_2-\gamma_3)C_E^0 \\ &= V\tilde{C}(T) - (V-\gamma_1)(\tilde{C}_C(T) - \tilde{C}_E(T)) + \gamma_2(\tilde{C}_B(T) - \tilde{C}_E(T)) \\ &= V\tilde{C}_E(T) + \gamma_1(\tilde{C}_C(T) - \tilde{C}_E(T)) + \gamma_2(\tilde{C}_B(T) - \tilde{C}_E(T)) \end{aligned}$$

and leads to

$$C_E^0 = \frac{1}{V - \gamma_1 - \gamma_2 - \gamma_3}\left[V\tilde{C}_E(T) - \frac{\gamma_1 G V_C}{(V_C + V_B + V_E)D_C} - \frac{\gamma_2 G V_B}{(V_C + V_B + V_E)D_B}\right] \tag{4.124}$$

Result: If T is so large that (4.120) can be assumed to hold, the solution $y_0 = (C_C^0, C_B^0, C_E^0)^T$ of (4.108) in step 3 is given by (4.122), (4.123), (4.124) where $\tilde{C}_E(T)$ is to be taken from (4.116). Finally, we will find out numerically how large D_B (≤ 25.28, see (4.103)) has to be chosen in order to guarantee (4.120) for $T = 2880$ and $t_d = 360$ $[min]$ (which were used for the two-compartment model already). We again choose V_C, V_B, V_E as in Section 4.4.2. The question then is how D_C and D_B with $D_C + D_B = 500$ $[ml/min]$ and $D_B \leq 25.28$ have to be chosen in order to make sure that (4.120) holds. For $D_B = 0.1, D_C = 499.9$ and $D_B = 1, D_C = 499$, respectively, we calculate by step 2 of the above procedure (using Euler's polygon method) the matrices

$$Y(2880) = \begin{pmatrix} 0.271658 & 0.006182 & 0.121257 \\ 0.053541 & 0.811637 & 0.024839 \\ 0.271629 & 0.006288 & 0.121244 \end{pmatrix}$$

and

$$Y(2880) = \begin{pmatrix} 0.266416 & 0.026670 & 0.119065 \\ 0.238152 & 0.141718 & 0.107808 \\ 0.266378 & 0.026823 & 0.119041 \end{pmatrix}$$

respectively, which have not yet the form of the right-hand side of (4.120). But if we choose $D_B = 10, D_C = 490$, then we obtain the matrix

$$Y(2880) = \begin{pmatrix} 0.270708 & 0.018321 & 0.121187 \\ 0.270708 & 0.018321 & 0.121187 \\ 0.270708 & 0.018321 & 0.121187 \end{pmatrix} \tag{4.125}$$

which is of the form of the right-hand side of (4.120). We also obtain

$$Y(360) = \begin{pmatrix} 0.280790 & 0.016226 & 0.125332 \\ 0.296510 & 0.087008 & 0.141592 \\ 0.248955 & 0.015355 & 0.111251 \end{pmatrix}$$

which, by virtue of (4.117), leads to

$$\gamma_1/V = 0.2707081,$$
$$\gamma_2/V = 0.0183211,$$
$$\gamma_3/V = 0.1211863$$

and shows good coincidence with (4.125). Therefore, for $D_B \geq 10$ $[ml/min]$ one can assume with sufficient accuracy that $Y(2880)$ (for $t_d = 360$ $[min]$) can be computed by calculating $Y(t_d)$ by step 2 of the above method and by putting

$$Y(2880) = \frac{1}{V} \begin{pmatrix} \gamma_1 & \gamma_2 & \gamma_3 \\ \gamma_1 & \gamma_2 & \gamma_3 \\ \gamma_1 & \gamma_2 & \gamma_3 \end{pmatrix}$$

where $\gamma_1, \gamma_2, \gamma_3$ are obtained from (4.117). This has the advantage that Euler's polygon method has only to be applied to the essentially smaller time interval $[0, t_d]$.

4.4.4 Numerical Results

We choose the same data as at the beginning of Section 4.4.2 and discuss the results for the values

$$\left.\begin{array}{llll} D_B = 0.1, & D_B = 1, & D_B = 10, & D_B = 20, \\ D_C = 499.9, & D_C = 499, & D_C = 490, & D_C = 480. \end{array}\right\} [ml/min]$$

As steplength for Euler's polygon method we have chosen $\Delta t = 1$ $[min]$ which guarantees a sufficiently high accuracy. For the initial values of the positive, absolutely continuous, and T-periodic solutions of (4.1) we obtain the following table:

D_B	$C_C(0) = C_C(T)$	$C_B(0) = C_B(T)$	$C_E(0) = C_E(T)$
0.1	1.14482	0.831434	1.15788
1	1.14421	0.923036	1.15770
10	1.13551	1.114731	1.14843
20	1.13309	1.129427	1.14628

This table shows that $C_B(0) = C_B(T)$ grows monotonically with growing D_B and gets closer to the corresponding value of $C_C(0) = C_C(T)$ which together with $C_E(0) = C_E(T)$ slightly decreases.
For $D_B = 25.28, D_C = 474.72$ (i.e., in the two-compartment model) one obtains the values

$$C_C(0) = C_B(0) = 1.13278, \quad C_E = 1.14611.$$

At the end of the dialysis (i.e., for $t = t_d = 360$ [min]) the following table for C_C, C_B, and C_E is obtained:

D_B	$C_C(t_d)$	$C_B(t_d)$	$C_E(t_d)$
0.1	0.527813	0.828088	0.480161
1	0.528787	0.869595	0.481363
10	0.534848	0.639948	0.487396
20	0.536505	0.553766	0.487695

References

[1] F. R. Gantmacher: Applications of Theory of Matrices. Interscience Publishers, New York - London - Sydney 1959.

[2] D. Klingelhöfer, G. F. Koch und W. Krabs: Mathematische Behandlung eines Modells der Haemodialyse. Math. Meth. Appl. Sc. 3 (1981), 393 - 404.

[3] W. Krabs: Über ein Drei-Kammer-Modell der Haemodialyse. ZAMM 63 (1983), 37-42.

[4] W. Krabs: Mathematische Modellierung. Verlag B.G. Teubner: Stuttgart 1997.

[5] J. A. Sargeant and F. A. Gotch: Principles and Biophysics of Dialysis. In: Jacobs C, Kjellstrand C, Koch K, Winchester J (eds): Replacement of Renal Function by Dialysis, ed 4. Dortdrecht. Kluwer Academic Publishers, 1996, 34-102.

A

Appendix

A.1 A Problem of Optimal Control

A.1.1 The Problem

We consider a function $f_0 : G \to \mathbb{R}$ and a vector function $f : G \to \mathbb{R}^n, f = (f_1, \ldots, f_n)^T$, on an open and connected domain $G \subseteq \mathbb{R} \times \mathbb{R}^n \times \mathbb{R}^r$.
Further we consider two vector functions $H_0, H_1 : V \to \mathbb{R}^n$ on an open and connected domain $V \subseteq \mathbb{R}^n$ such that

$$G \cap \mathbb{R} \times V \times \mathbb{R}^r \neq \varnothing \text{ (= empty set).}$$

We assume that f_0 and f as well as H_0 and H_1 are partially continuously differentiable with respect to all variables on G and V, respectively.
Finally, let $[t_0, t_1]$ with $t_0 < t_1$ a given time interval.
Then we consider the following

Problem of optimal control. Find two functions $u \in C([t_0, t_1], \mathbb{R}^r)$ and $y \in C^1([t_0, t_1], \mathbb{R}^n)$ such that

$$(t, y(t), u(t)) \in G \text{ for all } t \in [t_0, t_1], y(t_i) \in V \text{ for } i = 0, 1, \tag{A.1}$$

$$\dot{y}(t) = f(t, y(t), u(t)) \text{ for all } t \in [t_0, t_1], \tag{A.2}$$

$$H_0(y(t_0)) = H_1(y(t_1)) = \Theta_n \text{ (= zero vector in } \mathbb{R}^n\text{)} \tag{A.3}$$

and

$$J(y, u) = \int_{t_0}^{t_1} f_0(t, y(t), u(t))\, dt \tag{A.4}$$

is minimized.

The function $u \in C([t_0, t_1], \mathbb{R}^r)$ is interpreted as a control(function) and the function $y \in C^1([t_0, t_1], \mathbb{R}^n)$ as a state (function) of a process which develops according to (A.2) submitted to initial and end conditions (A.3). $J : C^1([t_0, t_1], \mathbb{R}^n) \times C([t_0, t_1], \mathbb{R}^r) \to \mathbb{R}$ is a cost functional in a general sense.

As an example let us consider the second problem of optimal control in Section 3.1. Here we have $G = \mathbb{R} \times \mathring{\mathbb{R}}_+ \times \mathring{\mathbb{R}}_+$ where

$$\mathring{\mathbb{R}}_+ = \{u \in \mathbb{R} \mid u > 0\} \quad (n = r = 1).$$

Further we have $V = \mathring{\mathbb{R}}_+$ and $H_0, H_1 : V \to \mathbb{R}$ are defined by

$$H_0(p) = p - p_0 \quad \text{and} \quad H_1(p) = p - p_T.$$

The functions $f_0 : G \to \mathbb{R}$ and $f_1 : G \to \mathbb{R}$ are defined by

$$f_0(t, p, v) = v$$

and

$$f_1(t, p, v) = [f(p) - g(v)]\, p,$$

respectively.
The time interval $[t_0, t_1]$ is given by $[0, T]$ and the cost functional J : $C^1([0, T], \mathbb{R}) \times C([0, T], \mathbb{R})$ reads

$$J(p, v) = \int_0^T v(t)\, dt.$$

All the assumptions made above are satisfied.

A.1.2 A Multiplier Rule

We start with defining a Lagrange function $L : G \times \mathbb{R}^n \times \mathbb{R}_+ \times \mathbb{R}^n \to \mathbb{R}$ by

$$L(t, y, u, \dot{y}, \lambda, p) = \lambda f_0(t, y, u) + p^T(\dot{y} - f(t, y, u)). \tag{A.5}$$

Then we have the following

Theorem A.1 *Let $(\hat{y}, \hat{u}) \in C^1([t_0, t_1], \mathbb{R}^n) \times C([t_0, t_1], \mathbb{R}^r)$ be a solution of the problem of optimal control. Then there exist multipliers $\hat{\lambda} \in \mathbb{R}_+$ and $\hat{l}_0, \hat{l}_1 \in \mathbb{R}^n$ which are not all vanishing and a function $\hat{p} \in C^1([t_0, t_1], \mathbb{R}^n)$ such that*
a) the Euler equation with respect to y given by

$$-\frac{d}{dt} L_{\dot{y}}\left(t, \hat{y}(t), \hat{u}(t), \dot{\hat{y}}(t), \hat{\lambda}, \hat{p}(t)\right) + L_y\left(t, \hat{y}(t), \hat{u}(t), \dot{\hat{y}}(t), \hat{\lambda}, \hat{p}(t)\right) = \ominus_n \quad \text{for all } t \in (t_0, t_1). \tag{A.6}$$

holds true together with the boundary conditions

$$\begin{aligned} L_{\dot{y}}\left(t_0, \hat{y}(t_0), \hat{u}(t_0), \dot{\hat{y}}(t_0), \hat{\lambda}, \hat{p}(t_0)\right) &= H_{0y}(\hat{y}(t_0))^T \hat{l}_0, \\ L_{\dot{y}}\left(t_1, \hat{y}(t_1), \hat{u}(t_1), \dot{\hat{y}}(t_1), \hat{\lambda}, \hat{p}(t_1)\right) &= -H_{1y}(\hat{y}(t_1))^T \hat{l}_1 \end{aligned} \tag{A.7}$$

where L_y and $L_{\dot{y}}$ is the gradient of L with respect to y and $\dot{y}$, respectively, and $H_{iy}(\hat{y}(t_i))$ is the Jacobi matrix of H_i at $\hat{y}(t_i)$, $i = 0, 1$, and
b) the Euler equation with respect to u given by

$$L_u\left(t, \hat{y}(t), \hat{u}(t), \dot{\hat{y}}(t), \hat{\lambda}, \hat{p}(t)\right) = \ominus_r \quad \text{for all } t \in (t_0, t_1) \tag{A.8}$$

holds true where L_u is the gradient of L with respect to u.
On using the definition (A.5) *the Euler equation* (A.6) *turns out to be equivalent to the so called adjoint equation*

$$\dot{\hat{p}}(t) = -f_y(t, \hat{y}(t), \hat{u}(t))^T p(t) + \hat{\lambda} f_{0y}(t, \hat{y}(t), \hat{u}(t))^T \quad \text{for all } t \in (t_0, t_1) \tag{A.9}$$

where $f_y(t, \hat{y}(t), \hat{u}(t))$ denotes the Jacobi matrix of f with respect to y and $f_{0y}(t, \hat{y}(t), \hat{u}(t))^T$ is the gradient of f_0 with respect to y at $(t, \hat{y}(t), \hat{u}(t))$, and the boundary conditions (A.7) *are equivalent with the so called transversality conditions*

$$\begin{aligned} \hat{p}(t_0) &= H_{0y}(\hat{y}(t_0))\, \hat{l}_0, \\ \hat{p}(t_1) &= -H_{1y}(\hat{y}(t_1))\, \hat{l}_1. \end{aligned} \tag{A.10}$$

Finally, the Euler equation (A.8) *reads*

$$f_u(t,\hat{y}(t),\hat{u}(t))^T\,\hat{p}(t) = \hat{\lambda} f_{0u}(t,\hat{y}(t),\hat{u}(t))^T \quad \textit{for all } t \in (t_0,t_1) \tag{A.11}$$

where $f_u(t,\hat{y}(t),\hat{u}(t))$ *is the Jacobi matrix of* f *with respect to* u *and* $f_{0u}(t,\hat{y}(t),\hat{u}(t))^T$ *is the gradient of* f_0 *with respect to* u *at* $(t,\hat{y}(t),\hat{u}(t))$.

Proof. For the proof of Theorem A.1 we refer to [1] in Chapter 3. □

An application of Theorem A.1 to the second problem of optimal control in Section 3.1 leads to the following statement: Let $(\hat{p},\hat{v}) \in C^1[0,T]\times C[0,T]$ an optimal pair. Then there exist multipliers $\hat{\lambda}_0 \geq 0, l_0, l_1 \in \mathbb{R}$ with $(\hat{\lambda}_0,\hat{l}_0,\hat{l}_1)^T \neq \Theta_3$ (= zero vector in $\mathbb{R}^3$) and a function $\hat{\lambda} \in C^1[0,T]$ such that

$$\dot{y}(t) = -\left[f'(\hat{p}(t)) + f(\hat{p}(t)) - g(\hat{v}(t)\right]\hat{\lambda(t)} \quad \text{for all } t \in (0,T), \tag{A.12}$$

$$\hat{\lambda}(0) = \hat{l}_0, \hat{\lambda}(T) = \hat{l}_1 \tag{A.13}$$

and

$$-g'(\hat{v}(t))\hat{p}(t)\hat{\lambda}(t) = \hat{\lambda}_0 \quad \text{for all } t \in (0,T). \tag{A.14}$$

If $\hat{\lambda}_0 = 0$ then it follows from (A.14) because of

$$g'(\hat{v}(t))\hat{p}(t) > 0 \quad \text{for all } t \in (0,T)$$

that

$$\hat{\lambda}(t) = 0 \quad \text{for all } t \in (0,T)$$

which implies $\hat{l}_0 = \hat{l}_1 = 0$, a contradiction to $(\hat{\lambda}_0,\hat{l}_0,\hat{l}_1) \neq \Theta_3$. Therefore it follows that $\hat{\lambda}_0 > 0$.

A.2 Existence of Positive Periodic Solutions in a General Diffusion Model

A.2.1 The Model

The two- and three-compartment models for the process of hemodialysis which are considered in Section 4 are special cases of a general diffusion model. This consists of n compartments which are pairwise separated from each other by porous or impermeable walls. In these compartments there is a substance in different, time-dependent concentrations $\chi_i = \chi_i(t), i = 1, \ldots, n, t \in \mathbb{R}$. In each compartment the substance is generated with a time-dependent generation rate $e_i = e_i(t)$ per time unit in the i-th compartment. From each compartment the substance is also extracted with a time-dependent extraction rate $c_i = c_i(t)$ per time unit. If V_i is the volume of the i-th compartment and

$$c_{ij} = c_{ji} \geq 0 \quad \text{for } i \neq j$$

the time-independent diffusion coefficient per time unit between the compartment i and j.

Then the temporal change of the i-th concentration χ_i for $i = 1, \ldots, n$ is described by the differential equation

$$V_i \, \dot{\chi}_i(t) = -\left(\sum_{\substack{j=1 \\ j \neq i}}^{n} c_{ij} + c_i(t) \right) \chi_i(t) + \sum_{\substack{j=1 \\ j \neq i}}^{n} c_{ij} \chi_i(t) + e_i(t). \tag{A.15}$$

With the definitions

$$a_{ij} = \frac{c_{ij}}{V_i} \qquad \text{for } i, j = 1, \ldots, n, i \neq j$$

$$d_i(t) = \frac{c_i(t)}{V_i}, \quad b_i(t) = \frac{e_i(t)}{V_i} \qquad \text{for } i = 1, \ldots, n$$

we can rewrite (A.15) in the form

$$\dot{\chi}(t) = -\left(\sum_{\substack{j=1 \\ j \neq i}}^{n} a_{ij} + d_i(t) \right) \chi_i(t) + \sum_{\substack{j=1 \\ j=i}}^{n} a_{ij}\chi_j(t) + b_i(t) \tag{A.16}$$

$$\text{for } i = 1, \ldots, n.$$

We assume that the functions $d_i = d_i(t)$ and $b_i = b_i(t)$ are periodic with period p, non-negative, piecewise continuous and continuous, respectively.

Now we put the question whether the system (A.16) has p-periodic solutions

$$\chi_i = \chi_t(t) \quad \text{for } i = 1, \ldots, n \text{ and } t \in \mathbb{R}.$$

which are positive and absolutely continuous. A positive answer to this question will be given in the next subsection under the natural assumptions

$$\sum_{i=1}^{n} d_i \not\equiv 0 \quad \text{and} \quad \sum_{i=1}^{n} b_i \not\equiv 0 \tag{A.17}$$

and a condition of non-decomposition which prevents the existence of isolated compartments so that the whole system is not divided into independent subsystems.

A.2.2 An Existence and Unicity Theorem

In vector and matrix form the system (A.16) for $i = 1, \ldots, n$ reads

$$\dot{\chi}(t) = A(t)\chi(t) + b(t), \quad t \in \mathbb{R}, \tag{A.18}$$

where

$$a_{ii}(t) = -\left(\sum_{\substack{j=1 \\ j \neq i}}^{n} a_{ij} + d_i(t) \right), \quad i = 1, \ldots, n.$$

For every choice of $t_0 \in \mathbb{R}$ and $\chi(t_0) \in \mathbb{R}^n$ the unique absolutely continuous solution of (A.18) is given by

$$\chi(t) = \overline{Y}(t)\{\chi(t_0) + \int_{t_0}^{t} \overline{Y}(s)^{-1} b(s)\, ds\}, \quad t \in \mathbb{R}, \tag{A.19}$$

with the unique $n \times n$-fundamental matrix function $\overline{Y} = \overline{Y}(t)$ which satisfies

$$\dot{\overline{Y}} = A(t)\overline{Y}, \quad t \in \mathbb{R},$$
$$\overline{Y}(t_0) = n \times n \text{ -unit matrix}$$

In addition to the assumption (A.17) we require the following non-decomposition condition: For every pair (i, j) with $i \neq j$ there exists a chain $i_1, i_2, \ldots, i_k$ of indices with

$$a_{i_1 i} > 0, a_{i_2 i_1} > 0, \ldots, a_{j i_k} > 0.$$

Then we can prove the

Theorem A.2 *Let* $t_0 \in \mathbb{R}$ *be given. Then for the solution* (A.19) *of* (A.18) *it is true that*

$$\chi(t_0) > \Theta_n \Rightarrow \chi(t) > \Theta_n \quad \textit{for all } t > t_0, \tag{A.20}$$
$$\chi(t_0) \geq \Theta_n \Rightarrow \chi(t) \geq \Theta_n \quad \textit{for all } t > t_0, \tag{A.21}$$
$$\chi(t_0) \geq \Theta_n,\ \chi(t_0) \neq \Theta_n \Rightarrow \chi(t) > \Theta_n \quad \textit{for all } t > t_0. \tag{A.22}$$

Proof. At first we have

$$\chi_i(t) = \exp\left(\int_{t_0}^{t} a_{ii}(s)\, ds\right) \cdot \left\{\chi_i(t_0) + \int_{t_0}^{t} \left(\sum_{\substack{j=1 \\ j\neq i}}^{n} a_{ij} x_j(s) + b_i(s)\right) \exp\left(-\int_{t_0}^{t} a_{ii}(\tau) d\tau\right) ds\right\} \tag{A.23}$$

for all $t \in \mathbb{R}$ and $i = 1, \ldots, n$.

If (A.20) were false, then there exists a minimal $t > t_0$ with $\chi_i(t) \leq 0$ for some $i \in \{1, \ldots, n\}$. Because of $\chi_i(t_0) > 0$ there exists some $\hat{t} \in [t_0.t]$ with

$$\sum_{\substack{j=1 \\ j \neq i}}^{n} a_{ij}\chi_j(\hat{t}) + b_i(\hat{t}) < 0$$

Because of $a_{ij} \geq 0$ for all $j \neq i$ and $b_i(\hat{t}) \geq 0$ this inequality can only hold true, if for at least one $j \neq i$ the inequality $a_{ij}\chi_j(\hat{t}) < 0$ holds true which is only true, if $a_{ij} > 0$ and $\chi_j(\hat{t}) < 0$. This contradicts the minimality of t. Hence (A.20) must be true.

The implication (A.21) follows from (A.20), since the solution (A.19) of (A.18) depends continuously on $\chi(t_0)$.
For the proof of (A.22) we choose some $i \in \{1, \ldots, n\}$ with $\chi_i(t_0) > 0$. Because of the continuity of $\chi_i = \chi_i(t)$ it follows that there exists some $\delta > 0$ such that

$$\chi_i(t) > 0 \quad \text{for all } t \in [t_0, t_0 + \delta].$$

From (A.23), (A.21) and the non-decomposition condition it follows for every $j \in \{1, \ldots, n\}$

$$\chi_{i_1}(t) \geq \exp\left(\int_{t_0}^{t} a_{ii}(s)\, ds\right) \times \int_{t_0}^{t} a_{i_1 i} x_s(s) \times \exp\left(-\int_{t_0}^{t} a_{ii}(\tau) d\tau\right) ds > 0$$

for all $t \in [t_0, t_0 + \delta]$ and by induction

$$\chi_j(t) > 0 \quad \text{for all } t \in [t_0, t_0 + \delta].$$

The rest of the implication (A.22) follows from (A.20) and $\chi(t_0 + \delta) > \Theta_n$. □

Conclusion. The fundamental matrix function $\overline{Y} = \overline{Y}(t)$ consists for all $t > t_0$ of positive elements.

A further conclusion of the non-decomposition condition is

Theorem A.3 *If for some $t > t_0$*

$$\sum_{i=1}^{n} d_i \not\equiv 0 \quad on\ [t_0, t].$$

then for the spectral radius $\rho(\overline{Y}(t))$ of $\overline{Y}(t)$ it follows that

$$\rho(\overline{Y}(t) < 1.$$

Proof. The matrix $A(t)$ in (A.18) has, for every $t \in \mathbb{R}$, the representation $A(t) = A - D(t)$ with a constant matrix A such that for $y = (1, \dots, 1)^T \in \mathbb{R}^n$ we have $Ay = \Theta_n$ and the diagonal matrix

$$D(t) = \begin{pmatrix} d_{ii}(t) & \dots & 0 \\ \vdots & \ddots & \vdots \\ 0 & \dots & d_{nn}(t) \end{pmatrix}.$$

From $Ay = \Theta_n$ it also follows that $A^T \tilde{y} = \Theta_n$ for some $\tilde{y} \in \mathbb{R}^n$ with $\tilde{y} \neq \Theta_n$ and therefore

$$\overline{Y}(t)^T \tilde{y} = \overline{Y}(t)^T \left(A^T - D(t)\right) \tilde{y} = -\overline{Y}(t)^T D(t) \tilde{y}. \tag{A.24}$$

We assert that $\tilde{y}$ can be choosen such that $\tilde{y} \geq \Theta_n$. If this were not the case, we can assume without loss of generality that there are indices k_1, k_2 with $1 \leq k_1 \leq k_2 \leq n$ such that

$$\tilde{y}_1 < 0, \dots, \tilde{y}_{k_1} < 0, \tilde{y}_{k_1+1} = \dots = \tilde{y}_{k_2-1} = 0, \tilde{y}_{k_2} > 0, \dots, \tilde{y}_n > 0.$$

This implies because of $A^T \tilde{y} = \Theta_n$ that

$$\left(\sum_{\substack{j=1 \\ j \neq i}}^{n} a_{ij}\right)(-\tilde{y}_i) + \sum_{j=k_2}^{n} a_{ji}\tilde{y}_j = \sum_{j=1}^{k_1} a_{ji}(-\tilde{y}_j) \quad \text{for } i = 1, \dots, k_1.$$

Adding all these equations leads to the equation

$$\sum_{i=1}^{k_1} \sum_{j=k_2}^{n} \left[a_{ij}(-\tilde{y}_i) + a_{ji}\tilde{y}_j \right] = 0$$

with $-\tilde{y}_i > 0$ for $i = 1, \ldots, k_1$ and $-\tilde{y}_j > 0$ for $j = k_2, \ldots, n$. This implies $a_{ij} = a_{ji} = 0$ for $i = 1, \ldots, k_1$ and $j = k_2, \ldots, n$ which contradicts the non-decomposition condition.
Therefore $\tilde{y} \leq \Theta_n, \tilde{y} \neq \Theta_n$ and it follows from (A.24) that

$$\overline{Y}^T \tilde{y} = \tilde{y} - \int_{t_0}^{t} \overline{Y}(s) D(s) \tilde{y} \, ds < \tilde{y}$$

which implies

$$\rho(\overline{Y}(t)) = \rho(\overline{Y}(t)^T) \leq \max_{i=1,\ldots,n} \frac{(\overline{Y}(t)^T \tilde{y})_i}{\tilde{y}_i} < 1$$

(see [1] in Chapter 4). □

Now we can formulate the existence and unicity statement as

Theorem A.4 *If $\sum_{i=1}^{n} d_i \not\equiv 0$ on $\mathbb{R}$, then there exists exactly one absolutely continuous and p-periodic solution of* (A.18). *This is of the form* (A.19) *for some $t_0 \in \mathbb{R}$ where*

$$\chi(t_0) = \left(E_n - \overline{Y}(t_0 + p) \right)^{-1} \tilde{\chi}(t_0 + p) \tag{A.25}$$

with $E_n = n \times n$-unit matrix and

$$\tilde{\chi}(t) = \overline{Y}(t) \int_{t_0}^{t} \overline{Y}(s)^{-1} b(s) \, ds. \tag{A.26}$$

If in addition $\sum_{i=1}^{n} b_i \not\equiv 0$ on $\mathbb{R}$, then

$$\chi(t) > \Theta_n \quad \textit{for all } t \in \mathbb{R}. \tag{A.27}$$

Proof. An absolutely continuous solution $\chi = \chi(t)$ of (A.18) is p-periodic, if and only if

$$\left(E_n - \overline{Y}(t_0 + p)\right)\chi(t_0) = \overline{Y}(t_0 + p) \int_{t_0}^{t_0+p} \overline{Y}(t)^{-1} b(t)\, dt.$$

Theorem A.3 implies the invertability of $E_n - \overline{Y}(t_0 + p)$ which implies the first assertion of Theorem A.4.

Because of $\tilde{\chi}(t_0) = \Theta_n$ it follows from the implication (A.21) that

$$\tilde{\chi}(t) \geq \Theta_0 \quad \text{for all } t > t_0.$$

If $\tilde{\chi}_i(t_0 + p) = 0$ for some $i \in \{1, \ldots, n\}$, then because of (A.22) it follows that

$$\tilde{\chi}(t) = \Theta_n \quad \text{for all } t \in [t_0, t_0 + p]$$

which implies

$$b \equiv \Theta_n \text{ on } [t_0, t_0 + p]$$

and contradicts $\sum_{i=1}^{n} b_i \not\equiv 0$ on $\mathbb{R}$. Therefore $\tilde{\chi}(t_0 + p) > \Theta_n$ and since $\left(E_n - \overline{Y}(t_0 + p)\right)^{-1}$ consists of positive elements (which can be derived from its representation as Neumann series), it follows from (A.25) that $\chi(t_0) > \Theta_n$ and from (A.20) that $\chi(t) > \Theta_n$ for all $t > t_0$. p-periodicity of $\chi = \chi(t)$ finally implies (A.27).
This concludes the proof of Theorem A.4. □

A.3 Asymptotic Stability of Fixed Points

We consider a continuous mapping $f : X \to X$ where X is a nonempty subset of $\mathbb{R}^n$. This mapping defines a time-discrete dynamical system which is given by the sequence $(f^n)\, n \in \mathbb{N}_0$ where

$$f^0(x) = x \text{ and } f^n(x) = \underbrace{f \circ f \circ \ldots \circ f}_{n\text{-times}}(x) \text{ for all } x \in X.$$

A point $x^* \in X$ is called a fixed point of f, if $f(x^*) = x^*$. A point $x^* \in X$ is called an attractor with respect to f, if there is a relatively open subset $U \subseteq X$ with $x^* \in X$ and

$$\lim_{n\to\infty} f^n(x) = x^* \text{ for all } x \in U.$$

A point $x^* \in X$ is called stable with respect to f, if for every relatively compact and relatively open subset $U \subseteq X$ with $x^* \in U$ there exists a relatively open subset $W \subseteq U$ with $x^* \in W$ and

$$f^n(x) \in U \text{ for all } x \in W \text{ and all } n \in \mathbb{N}_0.$$

A point $x^* \in X$ is called asymptotically stable with respect to f, if x^* is an attractor and stable with respect to f.
Now let $G \subseteq X$ be a nonempty subset.

Definition: A function $V : X \to \mathbb{R}$ is called a Lyapunov function with respect to f and G, if
(1) V is continuous on X,
(2) $V(f(x)) - V(x) \leq 0$ for all $x \in G$.

With these definitions we formulate the following

Theorem A.5 *Let $x^* \in X$ be a fixed point of f. Further let there be a relatively open subset $G \subseteq X$ with $x^* \in G$ and a Lyapunov function V with respect to f and G which is positive definite, i.e.,*

$$V(x) \geq 0 \textit{ for all } x \in G \textit{ and } (V(x) = 0 \Leftrightarrow x = x^*)$$

Then x^* is stable with respect to f.
If in addition

$$V(f(x)) < V(x) \textit{ for all } x \in G \textit{ with } x \neq x^*,$$

then x^* is asymptotically stable with respect to f.

Proof. Let $U \subseteq X$ be a relatively compact and relatively open subset of X with $x^*\backslash U$. If we put $U^* = U \cap G$, then U^* is also a relatively compact and relatively open subset of X with $x^* \in U^*$. Therefore there exists some $r > 0$ such that

$$\overline{B_r(x^*)} = \left\{x \in X | \left\|x - x^*\right\|_2 \leq r\right\} \subseteq U^*.$$

Since f is continuous in x^*, there exists some $s \in (0, r)$ with $f(B_s(x^*)) \subseteq B_r(x^*)$. If we put $B_{U^*} = B_s(x^*)$, then it follows that $f(B_{U^*}) \subseteq U^*$, B_{U^*} is open in X and $x^* \in B_{U^*}$.

Let us put

$$m = min\left\{V(x) | x \in \overline{U^*} \backslash B_{U^*}\right\}.$$

Because of $x^* \notin \overline{U^*} \backslash B_{U^*}$ it follows that $m > 0$. Now we define

$$W = \{x \in U^* | V(x) < m\}.$$

Then W is open in X and $x^* \in W \subseteq B_{U^*}$.

Now let $x \in W$ be chosen arbitrarily. Then it follows that $x \in B_{U^*}$ and hence $f(x) \in U^*$.

Further it follows that

$$V(f(x)) \leq V(x) < m, \text{ hence } f(x) \in W \subseteq B_{U^*}.$$

This implies $f^2(x) = f(f(x)) \in U^*$ and hence

$$V\left(f^2(x)\right) \leq V(f(x)) < m, \text{ hence } f^2(x) \in W.$$

By iteration we obtain

$$f^n(x) \in W \subseteq U^* \subseteq U \text{ for all } n \in \mathbb{N}_0.$$

This shows that x^* is stable with respect to f. From the construction of W we infer that W is relatively compact and open in X. Further we have $f(W) \subseteq W$, $x^* \in W$ and $\overline{W} \subseteq G$.

Now let $x \in W$ be chosen arbitrarily. We assume that $f^n(x) \nrightarrow x^*$. Then there exists a subsequence $(f^{n_k}(x))_{k \in \mathbb{N}_0}$ and some $\overline{x} \in \overline{W}$ with $\overline{x} \neq x^*$ and

$$\lim_{k\to\infty} f^{n_k}(x) = \overline{x}. \tag{A.28}$$

From $f^{n_k}(x) \in W$ for all $k \in \mathbb{N}_0$ it follows that $f(f^{n_k}(x)) \in W$ for all $k \in \mathbb{N}_0$ and hence

$$f(f^{n_k}(x)) \to f(\overline{x}) \in \overline{W} \subseteq G.$$

Therefore there exists a neighbourhood $B = \{z \in W \mid \|z - \overline{x}\|_2 \leq \epsilon\}$ of $\overline{x}$ with $x^* \notin B$ and $f(B) \subseteq G$. For every $z \in B$ we therefore have $V(f(z)) < V(z)$ and

$$q = \sup_{z\in B} \frac{V(f(z))}{V(z)} < 1, \text{ i.e., } V(f(z)) \leq qV(z) \text{ for all } z \in B.$$

Because of (A.28) there is some $k_0 \in \mathbb{N}_0$ with $f^{n_k}(x) \in B$ for all $k \geq k_0$. This implies for all $k \geq k_0$ because of $f^n(f^{n_k}(x)) \in W \subseteq G$ for all $n \in \mathbb{N}$

$$V(f^{n_{k+1}}(x)) = V\left(f^{n_{k+1}-n_k-1}\left(f^{n_k+1}(x)\right)\right) \leq V\left(f^{n_k+1}(x)\right) \leq qV(f^{n_k}(x))$$

and by iteration

$$V(f^{n_{k_0+l}}(x)) \leq q^l V(f^{n_{k_0}}(x)) \text{ for all } l \in \mathbb{N}.$$

This implies

$$\lim_{l\to\infty} V(f^{n_{k_0+l}}(x)) = 0$$

and hence

$$V(\overline{x}) = 0,$$

which is impossible because of $\overline{x} \neq x^*$. Therefore the above assumption $f^n(x) \nrightarrow x^*$ is false and it follows that

$$f^n(x) \to x^* \text{ for all } x \in W$$

which shows that x^* is an attractor with respect to f.
This concludes the proof of Theorem A.5. □

Index

Printing: Krips bv, Meppel
Binding: Stürtz, Würzburg

